M000158303

# CHARLES DARWIN'S RELIGIOUS VIEWS
## FROM CREATIONIST TO EVOLUTIONIST

# CHARLES DARWIN'S RELIGIOUS VIEWS

## FROM CREATIONIST TO EVOLUTIONIST

REVISED AND EXPANDED EDITION

David Herbert

**press**

www.joshuapress.com

*Published by*
Joshua Press, Guelph, Ontario, Canada
*Distributed by*
Sola Scriptura Ministries International
www.sola-scriptura.ca

Joshua Press 2009 (revised and expanded edition).
© 2009 David and Irene Herbert. All rights reserved. This book may not be reproduced, in whole or in part, without written permission from the publishers.

© 2009 Cover and book design by Janice Van Eck. Cover engraving by Thomas Johnson. Every effort has been made to use photographs and illustrations that are in the public domain.

*Library and Archives Canada Cataloguing in Publication*

Herbert, David, 1940-
    Charles Darwin's religious views : from creationist
to evolutionist / David Herbert. — Rev. ed.

Includes bibliographical references and index.
**ISBN 978-1-894400-30-5**

    1. Darwin, Charles, 1809-1882—Religion.
2. Evolution—Religious aspects. 3. Creationism. I. Title.

BL263.H47 2009       231.7'652       C2008-907831-4

*To Gary and Ilse Webb, two dear and dedicated Christian friends who have been a source of encouragement as we have grown together in understanding God, our Creator, and his marvellous creation.*

# Contents

# Foreword

C harles Darwin was by no means the first to advocate that the human race originated through some accidental 'First Cause.' An apostle of the widely disseminated theory of biogenesis, he has in the last few years become the 'poster boy' for a new and virulent form of atheism that sees all forms of religion as toxic, poisonous and detrimental to the planet: all forms of "fundamentalism" are to be denounced; the most noxious, of course, is Christianity.

As we enter 2009, staunch advocates of Darwinistic atheism are becoming increasingly vocal in their attempt to persuade the masses that God "probably does not exist." Millions of dollars are being raised for advertising campaigns that seek to instill doubt in time-honoured Christian belief. Part of the reason for this aggressive proselytizing is the fact that 2009 marks both the 150th anniversary of the publication of Darwin's *On the Origin of Species* and the 200th anniversary of his birth. The mania surrounding these milestones is somewhat reminiscent of the unbridled hysteria exhibited by some in the days leading up to Y2K.

Author David Herbert has done the world a tremendous service in writing this gem. It is an accurate description of the downward spiritual slide of one of the most controversial characters in the annals of world history. Darwin began his life as a nominal Christian. By the end of his journey to the legendary Galapagos Islands, Darwin had become completely convinced that time and chance could reasonably account for the varied and complex life systems found on planet earth. Fortunately, at the academy at least, Darwin is no longer taken seriously. Any reasonable and consistent scientist, regardless of religious stripe, is distancing himself or herself from the sheer logical folly that randomness and nothingness can explain life in any rational manner. Tragically, Darwin's theory is still the majority position taught in our State-run education systems. There is much work to be done here.

One of the most intriguing parts of this book, and a story that is seldom told, is the subplot involving the captain of HMS *Beagle*, Robert FitzRoy. He too began life as a nominal Christian. He too voyaged, with Darwin, on that fateful adventure to the far side of the world. No doubt he was plagued with Darwin's theories of randomness and nothingness. Unlike Darwin, FitzRoy became convinced that luck, chance and time could not satisfactorily substitute for the Creator God depicted in the first chapter of Genesis. Based on the evidence, FitzRoy was savingly converted by the gospel of Christ.

Our hope and prayer is that David's book will be part of the ongoing cultural debate regarding "origins." May it be used of God for the furtherance of his kingdom, the equipping of his saints and the glory of his risen Son. *Soli Deo gloria*.

January 28, 2009
*Heinz G. Dschankilic*
Executive Director, Sola Scriptura Ministries International

# Preface

There are some things that never seem to change. When I was writing the first edition of this book, I vividly recall that one of my students at Sir Frederick Banting Secondary School, in London, Ontario, had difficulty in believing that Charles Darwin (1809-1882) was religious. Almost twenty years later, when a visiting engineering professor was answering a questionnaire entitled, 'Charles Darwin's religious views,' his response was exactly the same as that of my former student. Since he had been raised in China as an atheist, he too believed that Darwin was not religious.[1]

Similarly, the attitude toward science and religion has not changed significantly over the years. Science—a tool, a powerful one at that, in explaining and utilizing the wonders of nature—is still seen as

---

1 Interview 225 (November 5, 2007). See the author's *Eternity Before Their Eyes* (London, Ontario: D & I Herbert, 2007), 73, in which an explanation is given for this response.

the modern panacea; it speaks with the voice of authority on every aspect of life. On the other hand, religion—and Christianity in particular—is viewed as inconsequential, or even irrelevant, to the pressing issues of life.

Peter Atkins, a leading Oxford University chemist and atheist, wrote in his article, "Science and Religion: Rack or Featherbed: The Uncomfortable Supremacy of Science," that science was undoubtedly a marvel of the human mind, "growing out of religion, but surpassing it as a means of understanding the world."[2] It goes without saying that this type of mindset, which trivializes religion and almost deifies science, would have difficulty in seeing Charles Darwin, an eminent scientist and evolutionist, as being religious.

This book—a spiritual biography—focuses primarily on the religious experiences of Charles Darwin's life. More specifically, its intent is to demonstrate how Darwin's rejection of the Bible led him to adopt the naturalistic assumptions that were foundational to his belief in evolutionism.[3]

A definition of religion would be appropriate at this point. Religion, as I view it, is an individual's response to these three eternal questions:

1. Where did we come from?
2. Why are we here?
3. Where are we going?

How one answers these key questions reveals a picture of one's religious beliefs. The main theme of this book is to discover how Charles Darwin approached these three questions—the basis for his religious portrait.

---

2   Karl W. Giberson and Donald A. Yerxa, *Species of Origins: America's Search for a Creation Story* (Lanham, MD: Rowman & Littlefield, 2002), 58.

3   I deliberately avoid using the terms "creation" and "evolution." Instead, I use "creationism" and "evolutionism," as they more accurately reflect the religious nature of this issue.

**GODDESS OF REASON** • During the French Revolution, the cult of reason and its accompanying goddess gained acceptance. The inversion of the crucifix below the young lady shows its contempt for Christianity.

Darwin was, as is everyone, a product of his time. His interactions, both with the nineteenth-century society into which he was born and with those individuals with whom he became intimate, became the determining factors in the 'evolution' of his belief system.

It should be remembered that "Darwin's century inherited the faith of the *philosophes*."[4] This faith of the *philosophes* or leading French intellectuals from the century before Darwin, such as Voltaire (1694-1778), Jean-Jacques Rousseau (1712-1778) and Denis Diderot (1713-1784), was based solely on human reasoning. These men embodied the spirit of the Enlightenment or the Age of Reason. To a man, they rejected the Bible and any supernatural revelation as being spurious and beyond the borders of rationality.

The *philosophes'* faith was called Deism. It maintained that the beauty and complexity of life demanded a Creator. This 'First Cause,'

---

4  J.R. Moore, "Charles Darwin and the Doctrine of Man" in *Evangelical Quarterly* 44 (1972), 199. See also the author's *The Faces of Origins: A Historical Survey of the Underlying Assumptions from the Early Church to Postmodernism* (London, Ontario: D & I Herbert, 2004), 69-83.

not in any way connected to the God of the Bible, created the world with its natural laws that operated independently of the Creator who then completely withdrew from his creation. If this Creator was not intimately involved in the affairs of the created order, some questioned whether he really existed. Thus, Deism naturally gave rise to atheism. Nevertheless, the faith of the *philosophes*, as will be shown, had a profound effect on Darwin's belief system.

No discussion of Darwin's religious life would be complete if one did not address his relationship with Christianity. Was he ever a Christian? If he was not, did he experience a so-called deathbed repentance?

This edition includes an appendix called, "Today's university students and Darwin." In 2007 and 2008, a number of my Christian friends assisted me in interviewing 350 students who were attending the University of Western Ontario (UWO), in London, Ontario. The students were asked to give their opinions concerning Charles Darwin's religious views and were invited to discuss their own religious perspectives. Many were pleased to have this platform to discuss the vital questions of life. The results of this survey are presented in Appendix 1.

Although I am presenting my research on Charles Darwin's relationship to Christianity, my own religious perspective will be continually before every reader. Consequently, I believe that it is very important for me to declare my particular bias. I still maintain a firm commitment to the Bible as the inerrant Word of the living God, and the only reliable source in answering the three eternal questions of life. I spent so much time over the years emphasizing to my students the importance of informing the reader of one's bias that I feel it is imperative to heed my own advice in this regard.

In his excellent biography of the life of Jonathan Edwards (1703–1758), Christian historian George Marsden stated that his purpose was to show how Edwards' religious life provided a framework in understanding him "as a person, a public figure and a thinker in his own place and time."[5] A similar historical methodology will be used to explore the religious life of Charles Darwin.

I am still convinced that every author is dependent upon a small coterie of highly talented and creative individuals whose assistance is invaluable. I want to acknowledge the following who read the entire manuscript: Dr. Jerry Bergman, Professor of Biology at Northwest State College in Ohio; Dr. Michael Fox, assistant professor at St. Peter's Seminary in London, Ontario; Dr. Wayne Frair, Emeritus Professor of Biology at The King's College in New York; Rev. Andrew MacLeod, historian and minister of Living Hope Community Church in St. Thomas, Ontario; Dr. Robert Newman, Emeritus Professor of New Testament and Christian Evidences at Biblical Theological Seminary in Pennsylvania; Ian Taylor, author of *In the Minds of Men*; and, Dr. William Phipps, Emeritus Professor of Religion and Philosophy at Davis and Elkins College in West Virginia, and author of *Darwin's Religious Odyssey*.

Above all, I want to thank my wife, Irene, for her tireless efforts in proofreading. Her commitment to the Lord Jesus Christ and her desire to see this book bring glory to him has been a source of inspiration to me.

---

5   George Marsden, *Jonathan Edwards: A Life* (New Haven: Yale University Press, 2003), 6.

# Chapter 1

# Religious heritage

C harles Robert Darwin was born on February 12, 1809 (the same day as Abraham Lincoln). Following the custom of the day, his parents, respected British gentry, had their boy baptized nine months later on November 17 at St. Chad's Anglican Church in Shrewsbury.[1]

With the exception of a fourteen-month truce, England had been at war with France for sixteen years. Britain, a war-weary nation, had weathered an aborted invasion by Napoleon Bonaparte (1769-1821), then master of Europe, and in 1809 was still reeling under the commercial blockade that both the British government and Napoleon had placed on each other. This relentless period of wars had driven up taxes and put countless families on church relief.[2]

---

1  A. Desmond and J. Moore, *Darwin: The Life of a Tormented Evolutionist* (New York: Warner, 1991), 12.

2  See J. Adamson, *English Education 1789-1902* (London: Cambridge University Press, 1930), 14. Adamson states that, as early as 1803, one out of every seven

In Shrewsbury, a town some 262 km (163 mi) northwest of London in the county of Shropshire, young Charles and his three older sisters and his brother, Erasmus (1804-1881), as descendents of the wealthy and influential Darwin-Wedgwood lineage, basked in the comfort and security of the leisured class. They were not faced with the stark reality of the economic hardship endured by the majority of the British people.

## DARWIN INFLUENCE

Robert Waring Darwin (1766-1848), Charles' father, was a gigantic man who stood 188 cm (6' 6") and weighed about 145 kg (230 lb.). In speaking of his father, Charles wrote: "His chief mental characteristics were his powers of observation and his sympathy, neither of which have I ever seen exceeded or even equalled."[3] As a medical doctor, these qualities were invaluable. His extraordinary ability to analyze human character and demonstrate a deep compassion for his patients made him "a first-class psychoanalyst."[4] Surgery he completely avoided as every operation then was done without the use of anesthetics.

Robert's domineering father, Erasmus Darwin (1731-1802), "one of the pre-eminent physicians of the eighteenth century,"[5] forced his son into the field of medicine to fill the void left by the death of Robert's older brother, Charles. His medical career, which had shown such promise, had ended abruptly. As a brilliant twenty-year-old student, he cut his finger while dissecting a child's brain. He later died as a result of the infection. Within a few weeks after this heart-rending tragedy, from which Erasmus never fully recovered, twelve-year old Robert was "being referred to as 'the young Doctor'."[6]

---

individuals was on church relief.

3   Charles Darwin, *The Autobiography of Charles Darwin (1809-1882)*, ed. Nora Barlow (London: Collins, 1958), 28. Hereafter it will be cited as *Autobiography*.

4   George A. Dorsey, *The Evolution of Charles Darwin* (N.Y.: Doubleday and Page, 1927), 26.

5   Ralph Colp, *To Be an Invalid* (Chicago: University of Chicago Press, 1977), 33.

6   B. and H. Wedgwood, *The Wedgwood Circle, 1730-1897* (Westfield, N.J.:

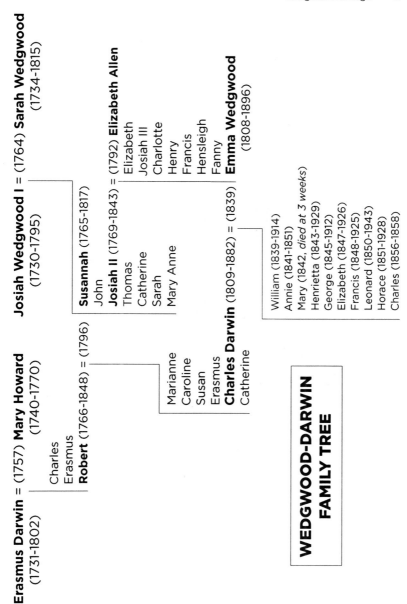

**WEDGWOOD-DARWIN FAMILY TREE**

Erasmus Darwin = (1757) Mary Howard
(1731-1802)                (1740-1770)

Charles
Erasmus
Robert (1766-1848) = (1796)

Josiah Wedgwood I = (1764) Sarah Wedgwood
(1730-1795)                (1734-1815)

Susannah (1765-1817)
John
Josiah II (1769-1843) = (1792) Elizabeth Allen
Thomas                   Elizabeth
Catherine                Josiah III
Sarah                    Charlotte
Mary Anne                Henry
                         Francis
                         Hensleigh
                         Fanny
                         Emma Wedgwood
                         (1808-1896)

Marianne
Caroline
Susan
Erasmus
Charles Darwin (1809-1882) = (1839)
Catherine

William (1839-1914)
Annie (1841-1851)
Mary (1842, *died at 3 weeks*)
Henrietta (1843-1929)
George (1845-1912)
Elizabeth (1847-1926)
Francis (1848-1925)
Leonard (1850-1943)
Horace (1851-1928)
Charles (1856-1858)

**FAMILY TIES** • The links between the Wedgwood and Darwin families are numerous. This family tree illustrates a few of them. Charles Darwin and his wife, Emma Wedgwood, were cousins.

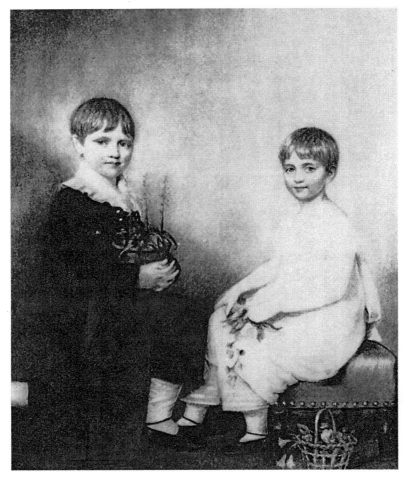

**THE TWO YOUNGEST DARWINS** • Charles at the age of 7, with his sister, Emily Catherine, 6, in 1816.

It is ironic that Robert Darwin, who had amassed such prosperity and public recognition through one of the largest medical practices outside of London, in reality, despised his profession. A

Eastview, 1980), 70. It should be noted that Erasmus Jr. (1759-1799), the second eldest son (see Wedgwood-Darwin family tree), who was never very highly regarded by his illustrious father, committed suicide at the age of forty.

skilled businessman, he was a major stockholder in the Trent and Mersey Canal which ran past the Wedgwood pottery factory in which he also had substantial investments.[7]

Dr. Erasmus Darwin's skill as a medical practitioner became well known and was duly recognized. King George III extended an invitation to him to be his personal physician but he declined. He was also an inventor, anatomist, poet and social activist against slavery.[8]

Erasmus also met regularly with a small group of leading scientists and industrial magnates. This group, never exceeding fourteen, included such men as Matthew Boulton (1728-1809), Josiah Wedgwood I (1730-1795) of 'Wedgwood pottery' fame, James Watt (1736-1819), who revolutionized the use of steam power and Joseph Priestley (1733-1804), Unitarian minister and noted chemist whose expertise was used in industrial research. They dined at each other's home on a Monday nearest the full moon and thus became known as the Lunar Society of Birmingham. Nevertheless, this closely-knit association of liberal-thinking savants became "the chief intellectual driving force behind the Industrial Revolution in England and hence the modern technological world."[9]

Their meetings were quite informal; they kept no records and never published any of their discussions on such diverse topics as electricity, optics, ceramics, steam-powered engines, canal building or their radical political views. Their support of the American colonists in their fight against the British monarchy put them at odds with the majority of their countrymen. Also, their sympathy towards the goals and aspirations of the French Revolution kindled the ire of British mainstream thought.

Erasmus Darwin, the first declared English evolutionist, best embodied the Enlightenment spirit as shown through his highly

---

7 Janet Browne, *Charles Darwin: Voyaging* (New York: Knopf, 1995), 8.

8 For a list of some seventy-five areas in which Erasmus Darwin did pioneer work, see D. King-Hele, *Doctor of Revolution: The Life and Genius of Erasmus Darwin* (London: Faber and Faber, 1977), 322.

9 King-Hele, *Doctor of Revolution*, 15.

controversial writings. The chapter "Of Generation" from his *Zoonomia; or, the Laws of Organic Life* (1794-1796) illustrates his rejection of the claims of the Bible as a supernatural authority.[10] In the following excerpt, Erasmus puts forth a number of naturalistic assumptions, including: eons of time, spontaneous generation, Deism and improvement without divine assistance. He writes,

> Would it be too bold to imagine, that in the great time since the earth began to exist, perhaps millions of ages before the commencement of the history of mankind, would it be too bold to imagine that all warm-blooded animals have arisen from one living filament [a simple cell], which THE GREAT FIRST CAUSE endued with animality, with the power of acquiring new parts, attended new propensities, directed by irritations, sensations, volitions and associations; and thus possessing the faculty of continuing to improve by its own inherent activity, and of delivering down those improvements by generation to its posterity, world without end.[11]

*Zoonomia* had a wide circulation[12] and was well received, possibly because Erasmus' medical expertise overshadowed the heretical chapter on evolutionism. But there was some opposition. For example, Samuel Taylor Coleridge (1772-1834), who wrote the famous poem, "The Rime of the Ancient Mariner," referred to *Zoonomia* as "the Orang Outang theology of the human race substituted for the first chapters of the Book of Genesis."[13]

---

10 See Norton Garfinkle, "Science and Religion in England 1790-1800," *Journal of the History of Ideas* 26 (1955), 377.

11 Erasmus Darwin, "Of Generation" in *Zoonomia; or, the Laws of Organic Life*, 2 vols. (London: J. Johnson, 1794), 1:505.

12 Norton Garfinkle states that within a decade there were four British editions of this book and two American. The book was also translated into French and German (Garfinkle, "Science and Religion in England 1790-1800," 380).

13 *Autobiography*, 151.

The greatest outcry came with the posthumous publication of *The Temple of Nature* in 1804. This lengthy poem outlined most clearly Erasmus Darwin's deistic beliefs on origins. His blatant disregard of God as Creator and his theory of evolutionary development of life from lower to higher forms was branded as rank atheism, especially as his writings became associated with the dastardly deeds of the French Revolution. Some years later Erasmus' renowned grandson, Charles, spoke with a great deal of pride that *Zoonomia*, in 1817, had been placed on the *Index Librorum Prohibitorum*—a list of books that Roman Catholics were forbidden to read.[14]

It is almost predictable that the skepticism of one generation becomes more extreme in the next. The deistic beliefs of Erasmus became atheistic in his son, Robert. Charles, Robert's son, was very much aware of the contempt that his father held for Christianity. Robert stated that he knew only three women who were skeptics, one being his sister-in-law Catherine (Kitty) Wedgwood (1774-1823), because "so clear-sighted a woman could not be a believer."[15] Only in light of the rigid Victorian stereotyping could one fully appreciate the outrageousness of his statement. Robert Stevens observes: "Skepticism, in short, was seen as a masculine characteristic—unsentimental, vigorous, self-reliant. Faith was seen as feminine—sentimental and dependent."[16]

Thus, in 1874, in responding to a questionnaire sent by Francis Galton (1822-1911) to many prominent English scientists in which they were to compare themselves with their fathers, Charles, then sixty-five, portrayed his father as being quite liberal for his time. The chart below shows the influence of Robert upon his son in the two areas pertinent to this study:

---

14 King-Hele, *Doctor of Revolution*, 241. In 1559, as a result of the Council of Trent, the first list was drawn up in reaction to the wide circulation of books written by the Protestant reformers. The *Index* was finally abolished in 1966.

15 *Autobiography*, 96.

16 L. Robert Stevens, *Charles Darwin* (Boston: Twayne, 1978), 16.

| Question | Yourself (Charles Darwin) | Your father (Robert Darwin) |
| --- | --- | --- |
| Religion | Nominally to Church of England | Nominally to Church of England |
| Independence | I think fairly independent; but I can give no instances. I gave up common religious belief almost indepentently from my own reflections. | Free-thinker in religious matters[17] |

Robert Darwin was as domineering over the lives of his two sons as his own father had been over his. Religiously, his two sons responded quite differently. Erasmus, the elder, tended to be an atheist like his father, though far more vocal. Charles, on the other hand, had a greater sensitivity toward the existence of God. But why? To answer that question, one must examine the influence of the Wedgwoods on Charles' life.

**WEDGWOOD INFLUENCE**

Susannah Wedgwood (1765-1817), Charles' mother, was the eldest and favourite daughter of Josiah Wedgwood I. His fine bone china company not only afforded him the title of Royal Potter of England but also gave him international acclaim and foreign markets for his famous pottery. Through the Lunar Society of Birmingham, he became a close friend of Joseph Priestley who had accepted a

---

17 Charles Darwin, *The Life and Letters of Charles Darwin*, ed. Francis Darwin, 2 vols. (New York: Basic Books, 1959), 2:356-357. Hereafter, it will be cited as *Life and Letters*.

**THE DARWIN-WEDGWOOD BELIEFS** • Charles Darwin's religious heritage on both sides of his ancestry.

call to be the minister of the Meeting Place, a Unitarian church, in Birmingham.

Priestley's research in gases had contributed a great deal to the discovery of oxygen. Recognizing the ability of this extraordinary chemist, Josiah supplied him with a well-furnished laboratory and utilized his discoveries to improve his own pottery production. "The Wedgwood firm even honoured their preacher by casting a medallion with his bust in bas-relief."[18]

Equally important to this pottery entrepreneur was Priestley's theological beliefs. Priestley had been raised and educated as a Dissenter and had become the foremost promoter of 'biblical Unitarianism' in Britain. His rejection of the teachings of the Church of England struck a chord with Josiah Wedgwood. Having long before distanced himself from the Anglican church, Wedgwood became an ally and supporter of Priestley's Unitarianism.

After only two years in his ministry at Birmingham, Priestley, a scientist-theologian, wrote *An History of the Corruptions of Christianity* in which he denounced the cardinal tenets of historical Christianity. He denied the deity of Jesus Christ (and the Holy Spirit), and Jesus' atonement for the sins of mankind. Jesus, wrote Priestley, a mere man, underwent inexpressible suffering "to exhibit a most perfect *example* of voluntary obedience to the will of God."[19]

In his autobiography, he stated that during the Sunday morning service, "I have also introduced the custom of expounding the

---

18 Desmond and Moore, *Darwin: The Life of a Tormented Evolutionist*, 8.

19 Joseph Priestley, *An History of the Corruptions of Christianity*, 2 vols. (Birmingham: Piercy and Jones, 1782), 1:180. Italics in the original.

**JOSEPH PRIESTLEY (1733-1804)** • A scholar and teacher, Priestley tried to unite Enlightenment rationalism and Christianity. His attempt was not well-received during his lifetime.

scriptures as I read them, which I had never done before, but which I would earnestly recommend to all ministers."[20] He believed in a providential God who directed the course of world history. To him, the Bible provided a standard of morality that had no equal. For

20 Joseph Priestley, *Autobiography of Joseph Priestley,* introduction by Jack Lindsay (Teaneck: Farleigh Dickenson University Press, 1970), 121.

those who had lived a good life, it gave hope beyond this present life. "Priestley firmly believed in the bodily resurrection, or more correctly, in a matter-spirit resurrection."[21]

Josiah was so impressed with this preacher-chemist that he appointed him teacher of his school at Etruria, Staffordshire, where his pottery factory was located. Both Charles Darwin's mother and his father were educated under Priestley. Besides being a noted author, having written thirty-four books during his lifetime,[22] Priestley was recognized as an innovative teacher. In 1761 he had written *The Rudiments of English Grammar* in which he separated English grammar from Latin grammar. This book enjoyed a great deal of popularity and was frequently consulted by English grammarians. Priestley also taught science, which was not part of the curriculum of the day, and provided scientific equipment for the students to use.

Being supportive of the French Revolution in 1791, some of Priestley's writings caused a serious rift between himself and the populace in Birmingham. Their disdain for him and his beliefs moved them to sack and burn down both the Meeting Place and Priestley's own house. Three years later, for the protection of his own life and that of his family, he was forced to emigrate to the United States where he authored *A General History of the Christian Church* and dedicated it to President Thomas Jefferson (1743-1826). Priestley died at the age of seventy.

Priestley's Unitarianism provided a solid foundation for "the three generations of intermarried Darwins and Wedgwoods."[23] Nevertheless, they never officially severed their ties with the Church of England. The ravages of rationalism had long eroded the theology of Anglicanism so that, in practice, it differed little from Unitarianism of the day.

---

21 Geoffrey Rowell, *Hell and the Victorians* (Oxford: Clarendon Press, 1974), 40.

22 Raymond J. Seeger, "Priestley, Nonconformist Minister" in *The Journal of the American Scientific Affiliation* 36 (December 1984), 241-242.

23 Desmond and Moore, *Darwin: The Life of a Tormented Evolutionist*, 8.

In the spring of 1817 at the age of eight, Charles' formal education began at the Unitarian Chapel day-school run by its minister, Rev. G. Case.[24] In his *Autobiography*, Charles spoke little about his time spent there. On Sundays, his mother took the family to the Chapel for services. After her death in July of that year Darwin's sisters, who were now grown women, became his surrogate mothers. "The 'sisterhood', as Charles dubbed them, were central to his early life."[25] They took him to the Anglican Church.

His sister Caroline (1800-1888), who was nine years older than Charles, became his 'governess'[26] after their mother's death. Charles' early moral and intellectual development was her responsibility. In a letter dated March 22, 1826, when Charles was at the University of Edinburgh, her concern for his spiritual well-being surfaced. She inquired of him:

> Dear Charles, I hope you read the bible and not because you think it wrong not to read it, but with the wish of learning there what is to feel and do to go to heaven after you die. I am sure I gain more by praying over a few verses than by reading many chapters. I suppose you do not feel prepared yet to take the sacrament.[27]

Her anxieties were somewhat placated when, in the next letter,[28] Charles reassured her that he was reading his Bible. To his inquiry about her favourite book, Caroline replied on April 11, 1826: "I agree with you in liking St. Johns the best of the Gospels. I am very fond of that short Epistle of St. James as well as St. Johns."[29] Frank

---

24 *Autobiography*, 22, footnote 1.

25 Browne, *Charles Darwin: Voyaging*, 11.

26 Charles Darwin, *The Correspondence of Charles Darwin*, ed. Frederick Burkhardt and Sydney Smith, 15 vols. (Cambridge: Cambridge University Press, 1985-forthcoming), 2:85. Hereafter, it will be cited as *Correspondence*.

27 *Correspondence*, 1:36.

28 *Correspondence*, 1:39. The letter was dated April 8, 1826.

29 *Correspondence*, 1:41.

**DR. SAMUEL BUTLER (1774-1839)** • Dr. Butler was the headmaster at
Shrewsbury Grammar School where Charles Darwin was a student from 1818
to 1825.

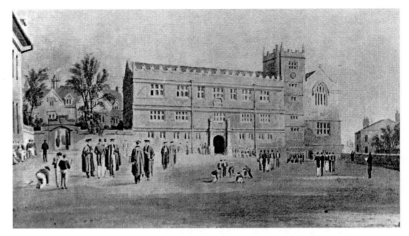

**SHREWSBURY GRAMMAR SCHOOL, 1833** • Charles and Erasmus spent their pre-university days here.

Burch Brown in his excellent book made the astute observation that Charles never mentioned "his apparent lack of readiness to take the sacrament."[30]

After one year at Rev. Case's school, Charles, in accordance with his father's high social standing, was sent to the elite Shrewsbury Grammar School. Dr. Samuel Butler (1774-1839), a graduate of Rugby and Cambridge, was the headmaster. During his thirty-eight-year tenure (1798-1836), he single-handedly restored the former prestige of this long-established school to that of Eton, Rugby and Harrow. The emphasis on studying and memorizing the Greek and Roman classics and New Testament was of little interest to young Charles. Nineteenth-century British educators believed that the classics fulfilled two purposes: first, to prepare their graduates for universities of either Oxford or Cambridge, controlled exclusively by the Anglican Church and, secondly, to mould their young men into "Christian gentlemen."[31]

---

30 Frank Burch Brown, *The Evolution of Darwin's Religious Views* (Macon, Georgia: Mercer University Press, 1986), 8.

31 Adamson, *English Education 1789-1902*, 59

A born scientist,[32] Charles Darwin, with his brother Erasmus, would escape from his classical studies to a make-shift laboratory in the garden tool shed. Dr. Butler, not appreciative of his intense interest in experimental science, rebuked Charles in front of his peers for wasting such time and added insult to injury by calling him a *poco curante*—a useless fellow.[33]

In relating his experiences at Shrewsbury, the boarding school which he attended for seven years, Darwin recalled in his *Autobiography* an incident which illustrated most cogently his awareness of God during his pre-university years. Since the school was only a kilometre or so from his house, Charles, having such strong family ties, frequently went home. Like most young boys, he put off returning to school until the last moment. If it appeared that he would be late for the school curfew, he wrote, "I prayed earnestly to God to help me and I well remember that I attributed my success to the prayers and not to my quick running and marvelled how, generally, I was aided."[34]

For many years, Maurice Mandelbaum's article written in 1959 was the accepted position on Darwin's religious views in his early years. He stated that prior to going to university, young Charles "was thoroughly orthodox."[35] However, in light of our previous discussion, we can conclude that Darwin was not committed in any way to the orthodox teaching of Christianity.

From the Darwin-Wedgwood heritage, Charles Darwin, prior to his university days, received a mixed message concerning a belief in God. It ranged from his father's impassioned skepticism to the ardent Unitarian faith of his mother and sisters. The Wedgwood tradition, supported in part by the ethical instruction at Shrewsbury, convinced young Charles that there was a Supreme

---

32 *Autobiography*, 22, 141.

33 *Autobiography*, 46. The students gave him the nickname, 'Gas' as he conducted experiments with numerous chemicals.

34 *Autobiography*, 25.

35 "Darwin's Religious Views" in *Journal of the History of Ideas* 19 (1958), 363.

Being and that the Bible was a book to be valued for its high moral principles; beyond that, there is little evidence of any deeper understanding.

## Chapter 2
# From medicine to theology

As a sixteen-year-old youth, Charles Darwin entered the University of Edinburgh on October 15, 1825. Known as the 'Athens of the North' and not bound by the *Thirty-Nine Articles*[1] as was Oxford and Cambridge, it became a haven for religious Dissenters, Independents and atheists. Edinburgh, the capital of Scotland, was influenced by the Enlightenment ideology of Continental Europe, especially France, and thus "evolutionary theories were in the air and embraced by many."[2]

---

1 This doctrinal statement was formulated during the reign of Elizabeth I (1533-1603) in 1563. It was the Confession of Faith for the Church of England. Evangelical in character, it attested to a Triune God, the authority of the Bible and the death and resurrection of Jesus Christ, the God-man.

2 Robert E. Kofal, "Charles Darwin: Influences on the Man, His Science, and His Theory," (1996), on-line at www.creationscienceoc.org/Articles/CharlesDarwin.html (accessed May 3, 2007).

Following family tradition[3] and also the wishes of his father, Charles enrolled in the Faculty of Medicine. Even though it was renowned for its medical teaching personnel, equipment and hospital facilities, his first year of studies proved to be a great disappointment. With the exception of chemistry, the lectures were extremely boring.[4] Even more shocking and unforgettable was the occasion that he visited the operating theatre:

> I saw two very bad operations, one on a child, but rushed away before they were completed. Nor did I ever attend again, for hardly any inducement would have been strong enough to make me do so; this being long before the blessed days of chloroform.[5]

Charles' most pleasurable moments were those spent with his brother, Erasmus, a second-year student, reading books from the university library on natural science. In a letter to his father, Charles wrote that he and his brother had attended the Church of Scotland. He was relieved that the sermon was only twenty minutes long. "I expected from Sir Walter Scott's account, a soul-cutting discourse of 2 hours and a half."[6]

In November 1826, Charles returned to Edinburgh alone as Erasmus left to study at Cambridge University. It was at this time that Charles began to give serious consideration about his educational future. Knowing full well that he would receive a substantial inheritance from his father, Charles realized that he really did not need to pursue a career. This belief, as recorded by himself, was

---

3 See Ralph Colp, *To Be an Invalid* (Chicago: University of Chicago Press, 1977), 3. Colp mentions the following four relatives of Darwin who were involved in medicine: his grandfather, Erasmus; his uncle, Erasmus; his father, Robert; and, his older brother, Erasmus.

4 *Autobiography*, 47; *Correspondence*, 1:25.

5 *Autobiography*, 48. Some twenty years later, in 1846, an anaesthetic was used by Robert Liston (1794-1847) for the first time.

6 *Correspondence*, 1:19.

**THE UNIVERSITY OF EDINBURGH, 1827**

sufficient to check any strenuous effort to learn medicine.[7] So, Charles began attending the lectures in geology and zoology instead of his medical classes. The family, and in particular Charles' father, was slowly coming to the realization that its youngest male member was not suited for the medical profession.

Meanwhile, Charles was encouraged to join the Plinean Society by Dr. Robert Edmond Grant (1793-1874), a highly respected zoologist and the Society's secretary. This group of students, numbering about twenty-five, had come together three years previously[8] to provide a forum in which they could discuss scientific matters. Their name was chosen in honour of the famous Roman naturalist, Pliny the Elder (A.D. *c.* 23-79). It soon became evident to Charles that these students were "fiery, freethinking democrats who demanded that science be based on physical causes, not super-

---

7  *Autobiography*, 46.

8  *Autobiography*, 50, footnote 1. The Plinean Society was formed in 1823 and disbanded in 1848.

natural forces."[9] For the rest of the school year, he attended these meetings.

Just three weeks or so after joining the Plinean Society, on December 16, 1826, Charles was afforded a rare opportunity. Accompanied by Dr. Grant, he attended a lecture given by the renowned American painter and naturalist, John James Audubon (1785-1851), at the Wernerian Society. The exquisite exhibition of some 400 paintings of birds set aflame a lifelong desire to study natural science. The decision to abandon his medical career was a *fait accompli*.

A very close relationship between Charles and Grant ensued to the point that, apart from Charles' father, no one during this period had such a profound influence on his life. Religiously speaking, this marine expert, like Darwin's father, was an atheist. He had studied and travelled extensively on continental Europe. But deep within his soul he was "a passionate Francophile."[10] Grant had the privilege of working with the prominent French scientists of the day but he had an unparalleled admiration for the aged Jean-Baptiste de Lamarck (1744-1829) who was then in his eighties.

Lamarck, a Deist and author of *Zoological Philosophy* (1809)—published the same year Darwin was born—clearly outlined the transmutational or evolutionary development of life. "The main purpose of the book was to put to rest the belief in the biblical concept of the fixity of kinds."[11] During one of the times that Grant and Charles were collecting marine specimens, Grant "burst forth in high admiration of Lamarck and his views on evolution."[12] Charles later wrote that he listened intently but made no comment. He was familiar with the concept of evolutionism as he had a few

---

9  A. Desmond and J. Moore, *Darwin: The Life of a Tormented Evolutionist* (New York: Warner, 1991), 31.

10  Desmond and Moore, *Darwin: The Life of a Tormented Evolutionist*, 34.

11  David Herbert, *The Faces of Origins: A Historical Survey of the Underlying Assumptions from the Early Church to Postmodernism* (London, Ontario: D & I Herbert, 2004), 77.

12  *Autobiography*, 49.

**JEAN-BAPTISTE DE LAMARCK (1744-1829)** • During his final years, Lamarck was not only blind but impoverished. He was the first French scientist to abandon the concept of the fixity of kinds and laid the foundation for evolutionism in Europe.

months before read his grandfather's book, *Zoonomia*, which he thoroughly enjoyed.[13]

Furthermore, to Charles' delight, his mentor must certainly have mentioned that he cited *Zoonomia* "in his doctoral thesis, and admitted that it opened his mind to some of 'the laws of organic life'."[14] Such praise lavished upon his celebrated grandfather undoubtedly left its mark on this impressionable student.

Knowing the mindset of the members of the Plinean Society, one is not surprised that Dr. Grant—a well-seasoned traveller and a rising intellectual—won the hearts of those science enthusiasts. His naturalistic, anti-biblical worldview only heightened their approval of him and his beliefs. For Charles Darwin, Grant was the first avowed evolutionist that he had personally encountered.

After leaving the University of Edinburgh in 1827, Dr. Grant was appointed the first Professor of Comparative Anatomy and Zoology at the University of London, known disparagingly as 'the godless college.' In 1836, he was made a Fellow of the Royal Society of London. But, by 1840, he was labelled an outcast for his adherence to Lamarckism. "Lamarckism had long been associated in the English minds with French atheism and social upheaval…it heralded racial extremism, if not revolution."[15] Grant's refusal to acknowledge a Creator within the scientific community committed to Natural Theology[16] gradually caused his total ostracism.

There seems to be little doubt that Grant made a lasting impression upon the teenaged Darwin. A naturalist of his stature, who blatantly and openly denigrated a biblical supernatural worldview

---

13 *Autobiography*, 49.

14 Desmond and Moore, *Darwin: The Life of a Tormented Evolutionist*, 40. Grant received his M.D. in 1814 from the University of Edinburgh.

15 A. Desmond, "Robert E. Grant: The Social Predicament of a Pre-Darwinian Transmutationist" in *Journal of the History of Biology* 17 (1984), 217f.

16 The design argument, a belief system, is based upon the concept that God could be seen and discovered in the mysteries and marvels of nature and should be clearly understood by everyone. Since Natural Theology was rationalistic by nature, it did not need the Scriptures.

in his classes, could not help but cause young Darwin to become sympathetic to a naturalistic perspective concerning origins. Furthermore, such a view was very similar to that held by his father, Robert.

So, one is not surprised that many years later in his *Autobiography*, when reflecting on Grant's influence, he admitted that "such views maintained and praised may have favoured my upholding them under a different form in my *Origin of Species*."[17] Being aware that his father "would leave him property enough to subsist on with some comfort,"[18] in the summer of 1827, Charles agreed with him that he should transfer to Cambridge and begin his studies for the Anglican ministry.

## AT CAMBRIDGE

Why did Darwin choose to become a minister? Why did he not become an army officer or a lawyer, as he had relatives who had successful careers in both? His father, knowing that his son loved natural science and that many ministers were also naturalists, realized the Anglican ministry offered his son "comparative security of position, opportunity for leisure, absence of any risk of failure."[19]

To a status-conscious family such as the Darwins, the ministry as a profession still opened doors to the cultured and highly respected circles of British society. Charles "needed a degree from an English university; family tradition was in this case applied and Cambridge was selected."[20] Neither Charles nor his father for one moment ever entertained the thought that 'a man of the cloth' was called by God to bring glory to his name through their service in the ministry.

---

17 *Autobiography*, 49.

18 *Autobiography*, 46.

19 James R. Moore, "Darwin of Down: The Evolutionist as Squarson-Naturalist," in David Kohn, ed., *The Darwinian Heritage* (Princeton: Princeton University Press, 1985), 442.

20 Peter Brent, *Charles Darwin: A Man of Enlarged Curiosity* (London: Heineman, 1981), 72.

Since Oxford and Cambridge were institutions controlled and operated by the Church of England, admittance to these universities was based on being a member. Upon graduation, each candidate had to sign a document which declared that he subscribed fully to the *Thirty-Nine Articles*.[21] Since Charles had been raised a Unitarian, he must have been troubled with the first article of the creed, 'Of Faith in the Holy Trinity'.[22] But, as always, expedience overruled conscience. Charles reasoned:

> As I did not in least doubt the strict and literal truth of every word in the Bible, I soon persuaded myself that our Creed must be fully accepted. It never struck me how illogical it was to say that I believed in what I could not understand.[23]

It is interesting to note that Darwin had to 'persuade' himself. With his strong Wedgwood heritage that greatly revered the Bible, this step of acceptance was easy; but he lacked a heart-based commitment. The Cambridge setting would do nothing to develop a deeper devotion to Christ and his Word. Here, Christianity was merely a veneer; one's commitment or lack of it mattered little. Chapel services were compulsory[24] but, in reality, they were meaningless.

Darwin's three years at Cambridge were pleasant, carefree and memorable ones. Since compulsory lectures were few in number and there were only periodic examinations, Charles was free to devote his energies to personal pursuits. Through his second cousin, William Darwin Fox (1815-1880) who had been sent to Cambridge

---

21 Janet Browne, *Charles Darwin: Voyaging* (New York: Knopf, 1995), 91. *Correspondence*, 1:122.

22 The last sentence of the first article reads: "And in unity of this Godhead there be three Persons, of one substance, power, and eternity; the Father, the Son, and the Holy Ghost."

23 *Autobiography*, 57.

24 J. Adamson, *English Education 1789-1902* (London: Cambridge University Press, 1930), 70. See *Life and Letters*, 2:141, where Darwin alludes to the ritualistic manner of the services.

to study theology, Darwin was introduced to the art of collecting beetles. Both of these aspiring clerics spent countless hours foraging the countryside in search of rare species. Charles experienced a delight reserved for only seasoned entomologists when he found "in Stephen's *Illustrations of British Insects* the magic words 'captured by C. Darwin Esq'."[25] The deep-seated love for natural science surfaced again.

University life also allowed this teenaged aristocrat to join "mainly well-to-do upper-middle class young men who came to school with dogs, guns and horses... attending lectures and their studies only when these were fun."[26] Hunting, riding and, most of all, partying marked this lifestyle. Darwin's own dog was named Sappho. These were British gentlemen; attending university was one of the entitlements that their social status afforded to them.

Charles joined the Gourmet or Glutton Club. Their gala events were "long-drawn-out dinners of an evening. They played cards. They actually sometimes drank too much."[27] One would hardly associate this type of conduct with a person entering the service of Christ. But it should be noted that more important than education gained at Cambridge were the social connections that one garnered; these privileged classmates would be the doctors, lawyers, politicians and clergymen of the future.

Biblical training was not considered to be an integral part of the theological curriculum.[28] According to an 1831 Cambridge University calendar, the examination for all B.A. candidates was divided into six parts:

---

25 *Autobiography*, 63.

26 Peter Nichols, *Evolution's Captain* (New York: Harper Collins, 2003), 102.

27 Geoffrey West, *Charles Darwin: A Portrait* (New Haven: Yale University Press, 1938), 71.

28 See Charles Darwin, *More Letters of Charles Darwin*, ed. Francis Darwin, 2 vols. (London: John Murray, 1923), 2:31. Hereafter, it will be cited as *More Letters*. Charles' lack of biblical knowledge is seen in a letter to L. Horner (Charles Lyell's father-in-law) on March 20, 1861, in which Darwin states he was amazed to learn that the year 4004 B.C. was not part of the original text.

1. Homer
2. Virgil
3. Euclid
4. Arithmetic and algebra
5. Paley's *Evidences of Christianity* and *Principles of Moral and Political Philosophy*
6. Locke's *An Essay Concerning Human Understanding* [29]

The two books written by the illustrious Anglican cleric, William Paley (1743-1805), formed the foundation upon which a rationalistic Christianity was built.

There is no better example than Darwin's own life. He stated that he spent so much time studying Paley's works in preparation for his final examination that "he could have written out the whole of the *Evidences* with perfect correctness."[30] This book, which presented an uncompromising defence of Christianity from a biblical and historical perspective, included a clear declaration concerning the resurrection of Jesus Christ, the linchpin of the gospel. Paley wrote:

> ...that the religion of Jesus was set up at Jerusalem, and set up with asserting, in the very place in which he had been buried, and a few days after he had been buried, his resurrection out of the grave, it is evident that, if his body could have been found, the Jews would have produced it, as the shortest and completest answer possible to the story. The attempt of the Apostles could not have survived this refutation a moment.[31]

Darwin, a ministerial student, approached *Evidences* from a purely scholastic viewpoint and its spiritual message failed to have any lasting impression on him.

---

29 *Correspondence*, 1:112, footnote 3.

30 *Autobiography*, 59.

31 William Paley, *A View of the Evidences of Christianity* (London: The Society for Promoting Christian Knowledge, 1871), 478.

**WILLIAM PALEY (1743-1805)** • His books, *A View of the Evidences of Christianity* and *Principles of Moral and Political Philosophy*, formed part of the core curriculum in Darwin's theological studies at Cambridge. Paley's last book, *Natural Theology*, also studied and admired by Darwin, was very popular during the nineteenth century.

Out of interest and because "he was charmed and convinced by his long argumentation,"[32] Charles read Paley's last book, *Natural Theology* (1802). This book very much represented the theological position of the day—Natural Theology. Its message, simply stated, is that the beauty and complexity seen in all living things demand a Designer, namely God. Using the human eye as the first illustration of divine ingenuity, Paley hoped that it—as Exhibit A— would slay the archenemy, atheism. Ironically, *Natural Theology* was written specifically to undermine the 'atheism' of Charles' grandfather, Erasmus.

God's goodness and beneficence were also central tenets of Natural Theology. Throughout his creation, God's inscrutable wisdom was evident everywhere; every facet of existence was perfectly and exquisitely fitted for the well-being and benefit of all life. "There was no room for maladapted structures or creatures, especially evolving ones, in [God's] good and perfectly ordered creation."[33] The inescapable conclusion was that God existed and that atheism was an absolute impossibility. Darwin concluded: "I do not think I hardly ever admired a book more than Paley's *Natural Theology*. I could almost formerly have said it by heart."[34]

Paley never referred to the Bible in his book but he assumed that it was the theological basis for his argument. But by the 1820s, the Bible, though revered, was not considered the ultimate authority. Nature itself was more than sufficient to speak on behalf of God.

As young Charles had done at Edinburgh, he sought out individuals at Cambridge who, like himself, loved natural science. William Darwin Fox, his cousin and an avid insect collector, invited Darwin to the home of Rev. John Stevens Henslow (1796-1861), Professor of Botany and rector of a large country parish.

---

32 *Autobiography*, 59.

33 D. Lamoureux, "Theological Insights from Charles Darwin," *Perspectives on Science and Christian Faith* (2004), 3 [available on-line at www.asa3.org/asa/pscf/2004/pscf3-04lamoureux.pdf (accessed March 5, 2007)].

34 *Correspondence*, 7:388.

Professor Henslow had a standing invitation for students and faculty alike to come to his home every Friday. "Serious talk about science was the norm, each guest mingling and conversing freely."[35] Charles attended regularly and soon developed such a strong friendship with his professor, that students referred to him as "the man who walks with Henslow."[36]

During his stay at Cambridge, no one had a greater influence on him than Henslow. Charles, who detested lectures of any kind, attended Henslow's botany classes and found them to be intellectually stimulating and "as clear as daylight."[37] In a letter to Fox when Henslow was tutoring him for his final examination, Darwin lavished his mentor with this praise: "The hour with him is the pleasantest in the whole day. I think he is quite the most perfect man I ever met with."[38]

His close association with Henslow caused Darwin's admiration of him as a dedicated teacher, scholar and clergyman to soar. In thinking about Henslow, Darwin later wrote:

> He was deeply religious and so orthodox that he told me one day, he should be grieved if a single word of the *Thirty-Nine Articles* were altered. His moral qualities were in every way admirable.[39]

But some modern Darwinian scholars are challenging Charles's assessment of the orthodoxy of his beloved professor. One letter, from Charles to William Darwin Fox on January 23, 1830, has greatly aroused their suspicions. Charles wrote: "I have heard men say that Henslow has some curious religious opinions; Have you?"[40]

---

35 Browne, *Charles Darwin: Voyaging*, 123.
36 *Autobiography*, 64.
37 *Life and Letters*, 1:162.
38 November 5, 1830, *Life and Letters*, 1:157f.
39 *Autobiography*, 64f.
40 *Correspondence*, 1:110.

Was it true that Henslow had some doubts about trinitarianism and was becoming sympathetic to the Unitarian position? Such an allegation has prompted two of his biographers to title one chapter, "A Liberal Churchman."[41]

There are definite indications that Professor Henslow had believed that the first chapters of Genesis must be interpreted figuratively. In 1823, he wrote a paper on the Genesis flood in which he suggested "a hypothesis of a non-miraculous cause of the deluge."[42] Possibly, he averred, a comet-astroid triggered the global disaster. No one could have doubted his unwavering loyalty to the Church of England but he was definitely not alone in his views on Genesis.

In 1831, the extent to which the Enlightenment worldview was taking hold in Britain can best be illustrated by the Rev. Adam Sedgwick (1785-1873). When he stepped down as President of the Geological Society, he publicly denied the biblical record of a global flood during Noah's lifetime:

> Having been myself a believer and to the best of my power, a propagator of what I now regard as *philosophic heresy*... I think it right, as one of my last acts before I quit this chair, thus publicly to read my recantation.[43]

A few months later, Henslow introduced Darwin to Sedgwick, Professor of Geology at Cambridge. This meeting proved to be very timely. Having successfully completed his B.A. examination at the end of January 1831 in which he stood tenth out of 178 candidates,[44] and having seriously questioned his suitability for the ministry,[45] Darwin began to study geology at Henslow's suggestion.

---

41 S.M. Walters and E.A. Stow, *Darwin's Mentor: John Stevens Henslow, 1796-1861* (Cambridge: Cambridge University Press, 2001), 155-173.

42 Walters and Stow, *Darwin's Mentor*, 162.

43 C.C. Gillespie, *Genesis and Geology* (New York: Harper and Row, 1951), 142. Italics are mine.

44 *Autobiography*, 59, footnote 1.

45 *Life and Letters*, 1:147.

**JOHN STEVENS HENSLOW (1796-1861)** • Henslow was Darwin's favourite professor at Cambridge and exerted considerable influence on him.

So, when Sedgwick offered to take Darwin on a three-week geological trip to North Wales in the summer of that year, he gladly accepted. To learn the fundamentals of geology for his upcoming trip to the Canary Islands was to be his goal. It goes without saying that Sedgwick, who discarded the biblical view of a global flood that had long been the basis for geology, would have instructed Darwin in his newly-acquired naturalistic beliefs of uniformitarianism.[46]

Both Henslow and Sedgwick, as eminent scientists in their respective fields, grounded their devoted and aspiring naturalist in the tenets of Natural Theology. Much to their chagrin, Darwin would later adopt naturalistic assumptions that would not only deny the necessity of the Bible but also the constant involvement of a sovereign God within his creation. Three decades later, when Darwin published his book, *On the Origin of Species*, Adam Sedgwick became his most outspoken critic.

Before leaving Darwin's college days, we should consider a letter he wrote to William Darwin Fox on April 23, 1829. Five days earlier, tragedy had struck the home of his cousin: Fox's sister had died. In an effort to console his intimate friend, Darwin wrote these touching words:

> I feel most sincerely and deeply for you and your family.
> But at the same time, as far as anyone can, by his own good
> principles and religion be supported under such a misfor-
> tune you, I am assured, well know where to look for such
> support. And after the pure and holy a comfort as the Bible,
> I am equally assured how useless the sympathy of all friends
> must appear, although it be as heartfelt and sincere, as I
> hope you believe me capable of feeling.[47]

Tragedy often tests one's spiritual resources. At this time Darwin, still clinging to his Wedgwood spiritual heritage, could convey with utter confidence the strength and solace offered by the Scriptures to a grieving fellow student.

As Charles faced the fall of 1831, he never imagined that he would be approached to take part in a scientific endeavour of a

---

46 Uniformitarianism will be discussed at length in the next chapter.
47 *Correspondence*, 1:84.

much greater scope than the one he was presently planning for the Canary Islands. This experience would cause him to question his commitment to Progressive Creationism[48] and provide a new set of assumptions that would allow him to drift into the faith of Naturalism.

---

48 Lamoureux, "Theological Insights from Charles Darwin," 3. At that time, Darwin maintained that species were immutable or fixed and that God had intervened at different times as witnessed in the geological record.

## Chapter 3
# "A zealous disciple of Lyell"[1]

**N**ames like iguanodon, pterodactyl and dinosaur[2] became part of the English language during the early decades of the 1800s. Geology, the newest science, had captured the general public's imagination; amateur geologists sprang up everywhere. Going to the countryside to 'geologize' was, according to Victorian standards, a scientific pastime suitable even for women. Between 1820 and 1840, more books on geology were sold than English novels.[3] It was in 1807 that the first geological society was formed in Britain. This scientific organization received its Royal Charter from George IV in 1825.

---

1  *Life and Letters*, 1:234.

2  In 1841, Richard Owen (1804-1892) introduced his newly-coined word, dinosaur—from the Greek *deinos,* meaning terrible, and *sauros*, meaning lizard—to the British Association for the Advancement of Science.

3  Owen Chadwick, *The Victorian Church*, 2 vols. (New York: Oxford University Press, 1970), 1:558.

Traditionally, a biblical supernatural worldview was accepted in understanding the formation of the earth's surface. The strata was believed to have been laid down during the global flood of Noah's day as outlined in Genesis 6 to 9. But during this flurry of geological activity during the eighteenth century, the rules for interpreting the earth's surface had changed.

*The Theory of the Earth* (1795), the earliest geological treatise written by the Scottish author James Hutton (1726-1797), introduced the concept of unformitarianism. This method of interpreting the geological strata assumed that changes occurred over long periods of time by such natural agents as erosion, sedimentation, glaciation and volcanization. In other words, "there was no need to invoke miraculous catastrophes; all had been accomplished by natural causes."[4] The somewhat constant or uniform rate at which these geomorphic forces have altered the face of the earth became the foundational basis for the theory of uniformitarianism.

Hutton, following the Enlightenment spirit, was committed to the naturalistic worldview. As a Deist, he believed in a Divine Being who had created the world but had left its operation to natural causes eons before. Consequently, he discounted the supernatural biblical flood. It is significant to note that the Noahic flood was never disproved nor refuted, just flatly rejected.

The acceptance of the earth's history extending back eons of time found an unusual ally at the beginning of the nineteenth century. He was the Rev. Dr. Thomas Chalmers (1780-1847), an evangelical Scottish divine. He was greatly influenced by John Playfair (1748-1819), a professor of mathematics at the University of Edinburgh and an ordained minister of the Church of Scotland. Playfair, an ardent defender of Hutton's theory of uniformitarianism, helped to popularize it through his widely-read book, *Illustrations of the Huttonian Theory of the Earth* (1802).

After being converted through the writings of Jonathan Edwards,

---

4   Tess Cosslett, ed., *Science and Religion in the 19th Century* (Cambridge: Cambridge University Press, 1984), 6.

**JAMES HUTTON (1726-1797)** • A physician by training, Hutton turned his attention to geology and became known as the 'father of uniformitarianism.'

Chalmers became a dynamic proclaimer of the Word of God in both Scotland and England. It was said that at his church in Glasgow, "his sermons on Thursday afternoon brought local businesses to a standstill."[5] Unfortunately, in his zeal to accommodate Hutton's long ages within the Bible, Chalmers originated the gap or ruin-restoration theory. Diagrammatically, the gap theory is as follows:

---

5  David N. Livingstone, *Darwin's Forgotten Defenders* (Grand Rapids: Eerdmans, 1987), 8.

**THOMAS CHALMERS (1780-1847)** • Trained in mathematics, Chalmers was later ordained and pastored a large congregation at the Tron in Glasgow. He became the first moderator of the Free Church of Scotland.

*Genesis 1:1 "In the beginning, God created
the heavens and the earth."*

---

Gap extending over billions of years

---

*Genesis 1:2 "The earth was
without form and void…"*

As early as 1814, Dr. Chalmers presented this harmonization of
Hutton's uniformitarian theory and the Genesis account. As an
influential cleric and leading intellectual, he did much to undermine the traditional concept of a young earth.[6]

But 'the high priest of uniformitarianism' was to be Charles
Lyell (1797-1875). This lawyer-turned-geologist would capitalize
on the prevalent Enlightenment spirit. He was able to popularize
Hutton's theory of the gradual formation of the earth's surface
over eons of time through his much-circulated, three-volumed
book, *Principles of Geology* (1830-1833) and, more importantly,
through his contacts within the learned societies, especially the
Geological Society of London. From 1825 to 1837, he was its
president. It was the *Principles of Geology* that "administered the
*coup de grâce* to the deluge."[7]

Lyell was keenly aware of obtaining and maintaining public
opinion in order to have uniformitarianism accepted. He wrote to
George Scrope (1797-1876), the reviewer of his first volume of
*Principles* in the prestigious *Quarterly Review*, on June 1830:

If Murray [the publisher] has to push my volumes and you

---

6 The gap theory was popularized in North America through the footnotes of
the 1909 *Scofield Reference Bible*. Today, this reference has been deleted. A capable
theological and historical refutation can be found in Weston W. Fields' *Unformed
and Unfilled: A Critique of the Gap Theory* (Nutley, New Jersey: Presbyterian and
Reformed Publishing, 1976).

7 C.C. Gillespie, *Genesis and Geology* (New York: Harper and Row, 1951), 140.

wield the geology of the *Quarterly Review*, we shall be able
in a short time to work an entire change in public opinion.[8]

Raised in the Anglican Church, which allowed him to attend
Oxford University, Lyell later converted to Unitarianism (or Deism)
and worshipped regularly at the Little Portland Street Unitarian
Chapel in London.[9] Spurred on by his new faith commitment, Lyell
dedicated himself to the task of destroying the credibility of the biblical flood as a means of explaining the geological formations of the
earth. In other words, so long as the earth was seen as being a few
thousand years old, supernatural intervention was indeed possible,
but if the age of the earth could be reckoned to be millions of years,
he reasoned that no supernatural involvement was necessary. The
title page of his book exemplifies his intentions. It reads, "*Principles
of Geology*, Being an attempt to explain the former changes of the
Earth's surface by reference to the causes *now* in operation."[10]

The long-accepted global deluge, from Lyell's perspective, was
totally mythical. As we have noted previously, church leaders, even
an evangelical such as Dr. Chalmers, inadvertently aided his cause.
Their deliberate harmonization of long ages into the biblical text
set up this inevitable sequence: "First to go was Genesis-time: next
to go a universal flood."[11]

The immediate success of *Principles of Geology*, which went
through eleven editions during Lyell's lifetime, can be seen in the
acceptance of uniformitarianism by his peers in the Geological
Society of London. Sandra Herbert mentions that by 1834 the
majority of the members had jettisoned the biblical supernatural
explanation in favour of a naturalistic uniformitarian view.[12] The

---

8   Gillespie, *Genesis and Geology*, 134.

9   Sandra Herbert, *Charles Darwin, Geologist* (Ithaca, New York: Cornell University Press, 2005), 187.

10  Italics mine.

11  Chadwick, *The Victorian Church*, 1:559.

12  "Darwin the Young Geologist," in David Kohn, ed., *The Darwinian Heritage*
(Princeton: Princeton University Press, 1985), 489.

**CHARLES LYELL (1797-1875)** • Lyell popularized James Hutton's uniformitarianism but never fully accepted Darwinian evolutionism.

stage was now set for Lyell's most zealous disciple, Charles Darwin, to reinterpret natural science into a belief system. C.C. Gillespie's pithy statement correctly catches the relationship of Lyell's work to Darwin when he states: "Uniformitarianism in geology seemed to cry out for evolutionism in biology."[13]

13 Gillespie, *Genesis and Geology*, 131.

## DARWIN'S RELIGIOUS VIEWS WHILE ON HMS *BEAGLE*

"With the end of the Napoleonic Wars came an enormous expansion of world trade, the cornerstone of which was the *Pax Britannica*."[14] To ensure the operational safety of the trading vessels on the high seas, the British government ordered that accurate maps and charts be provided to all their naval captains.

Captain Robert FitzRoy (1805-1865), "only twenty-six himself, wanted a young companion, a well-bred 'gentleman' who could relieve the isolation of command, someone to share the captain's table."[15] It seems that Captain FitzRoy had contacted the British Admiralty to have a 'dining companion' accompany him while he surveyed the coasts of South America and the islands of the Pacific Ocean.

"If you can find any man of common sense who advises you to go, I will give my consent."[16] These words of Charles' father opened the door for him to approach his maternal uncle, Josiah Wedgwood II, in order that he might become the naturalist companion to Captain FitzRoy aboard HMS *Beagle*. This trip offered Darwin the opportunity of circumnavigating the world.

Through George Peacock (1791-1858), Leonard Jenyns (1800-1893), a Cambridge mathematics professor, and Professor John Henslow had been asked to go—but both declined. Henslow, who was aware of Darwin's desire to travel abroad as a naturalist, wrote this highly complimentary letter on August 24, 1831 to his esteemed student:

I consider you to be the best qualified person of who (*sic*)

---

14 Keith S. Thomson, *HMS Beagle: The Story of Darwin's Ship* (New York: Norton, 1995), 48. Napoleon had been defeated in Belgium on June 18, 1815 at the Battle of Waterloo. *Pax Britannica* is Latin for "the British peace" (modelled after the phrase *Pax Romana*) and refers to the period after Napoleon's defeat when the British enjoyed naval superiority and pursued the expansion of their empire.

15 A. Desmond and J. Moore, *Darwin: The Life of a Tormented Evolutionist* (New York: Warner, 1991), 101.

16 *Autobiography*, 71.

I know is likely to undertake such a situation—I state this not on the supposition of you being a *finished* naturalist, but as amply qualified for collecting, observing, and noting any thing worthy to be noted in Natural History. Peacock has the appointment at his disposal and if he cannot find a man willing to take the office, the opportunity will be probably lost.[17]

Robert Darwin's initial reaction to his son's wanting to be part of this purposeless and dangerous lark was that it would mar his standing as a respected clergyman. But after reading a letter dated August 31, 1831, from his brother-in-law Josiah Wedgwood II, which answered all eight of his objections, he gladly gave his permission. In his *Autobiography*, when reflecting on the importance of the *Beagle* trip, Darwin remarked: "It determined my whole career... I owe to the voyage the first real training or education of my mind."[18]

So, after a few months delay, HMS *Beagle*,[19] a 242-ton brig-sloop with a crew of seventy-four, set sail on December 27, 1831 from Plymouth on its 64,000 km (40,000 mi.) journey around the world. The ship was 27.5 m (90') long—"no longer than the distance between the bases on a baseball field"[20]—and 7.5 m (25') wide. On the official ship's list, Charles was recorded as a 'supernumerary.'[21] This status also afforded him complete independence throughout the whole trip; he could come and go as he pleased. But equally important was that all the specimens that he found on the voyage were his and they were all sent to his loyal professor, John Henslow, who housed them until his return.

---

17 *Correspondence*, 1:128f. Italics are in the original.

18 *Autobiography*, 76-77.

19 HMS *Beagle* was commissioned in 1820 and decommissioned from the Royal Navy in 1845. It was used for three survey voyages.

20 Thomson, HMS *Beagle: The Story of Darwin's Ship*, 133.

21 Herbert, *Charles Darwin, Geologist*, 22. See *Correspondence*, 1:549, for the entire list of supernumeraries.

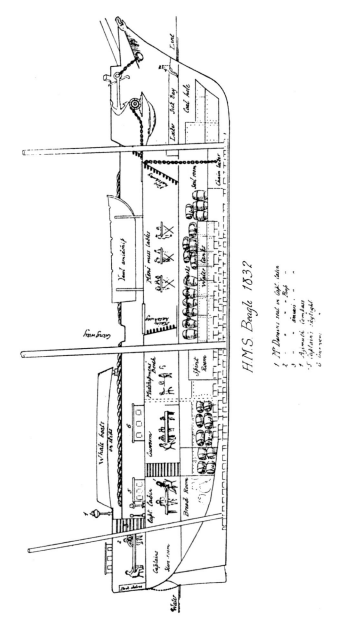

**HMS *BEAGLE*** • At the age of eighty in 1890, Philip King, a former crew member, drew this sketch of HMS *Beagle*.

Every Sunday, while aboard ship, Charles attended the compulsory worship service conducted by Captain FitzRoy. "The *Beagle* artist, Augustus Earle [1793-1838], painted a detailed scene of the ship's company participating in a typical service."[22] In his *Beagle* diary, Darwin mentions that, on the Lord's Day, he would read from his Greek New Testament.[23]

On Sunday, November 4, 1832, Darwin and Robert Hamond (1809-1883), who had just joined the *Beagle* on July 31 and had become a good friend, visited the churches in Buenos Aires, Argentina. They were impressed with the fervency of devotion and the equality among the Catholic worshippers. "The Spanish lady with her brilliant shawl knelt by the side of her black servant in the open aisle."[24]

Darwin and Hamond approached a naval chaplain and asked him to administer communion for them before entering the dangerous waters of Tierra del Fuego at the tip of South America. He refused but suggested that they return with other members of the crew. These two shipmates never did. Instead, they "rejoined the *Beagle*, thrown back on FitzRoy's seamanship and other means of grace."[25]

In his *Autobiography*, Darwin referred to himself as being quite orthodox (or biblically Unitarian). During a discussion with the officers of the *Beagle*, he quoted the Bible "as unanswerable authority on some point of morality."[26] These men chided him for resorting to the Scriptures.

The breathtaking beauty of the tropics was something that this unseasoned traveller never forgot. Once in the stillness and serenity

---

22 William E. Phipps, *Darwin's Religious Odyssey* (Harrisburg: Trinity Press International, 2002), 17.

23 Charles Darwin, *Diary of the Voyage of H.M.S. Beagle*, ed. Nora Barlow, vol. 1 in *The Works of Charles Darwin,* Paul H. Barrett and R.B. Freeman, eds. (London: Pickering, 1986), 1:15. Hereafter, it will be cited as *Diary*.

24 *Diary*, 1:102.

25 Desmond and Moore, *Darwin: The Life of a Tormented Evolutionist*, 130.

26 *Autobiography*, 85.

of a Brazilian forest, he experienced feelings of "wonder, admiration and devotion which fill and elevate the mind."[27] It was here that he realized that there was something greater than mankind.

On November 15, 1835, "at daylight, Tahiti, an island which must for ever remain as classical to the Voyager in the South Sea, was in view."[28] Before arriving, Darwin had read three books which were part of the library on the *Beagle* concerning life on Tahiti.[29] The three authors' attitudes toward the Christian missionaries in this tropical paradise varied greatly—most critical was Otto von Kotzebue (1787-1846). This German explorer travelled around the world with the Russian Navy, visiting Tahiti in 1824, just ten years before Darwin. He acknowledged that Christianity had "abolished heathen superstitions and irrational worship but it has introduced new errors in its stead."[30] He felt that the missionaries were not properly trained and were enforcing an unrealistic Christianity upon the Tahitians. Von Kotzebue wrote:

> A religion which consists in the eternal repetition of prescribed prayers, which forbids every innocent pleasure, and cramps or annihilates every mental power, is a libel on the Divine Founder of Christianity, the benign Friend of humankind.[31]

During his ten-day stay, Charles' assessment was radically different from that of von Kotzebue. He and the *Beagle* crew enjoyed the beauty and the hospitality of this enchanted island. In a letter to his sister Caroline, he wrote:"The missionaries have done much

---

27 *Autobiography*, 91.

28 *Diary*, 1:313.

29 *Diary*, 1:323. See "Appendix IV: The books on board the *Beagle*" in *Correspondence*, 1:553-566. The ship's library consisted of some 245 books.

30 Otto von Kotzebue, *A New Voyage Round the World in the Years: 1823-1826*, 2 vols. (N. Israel/Da Capo: Amsterdam/New York, 1967), 2:168. See also H.E.L. Mellersh, *FitzRoy of the Beagle* (London: Rupert Hart-Davis, 1968), 151-152.

31 von Kotzebue, *A New Voyage Round the World*, 2:168.

in improving their moral character and still more in teaching them the arts of civilization."[32]

The prayer life of the Tahitians caught Darwin's attention. He noted that they regularly gave thanks before each meal. On one of his treks into the mountains he witnessed firsthand that, prior to bedding down for the night, his Tahitian guide fell on his knees and "prayed as a Christian should do, with fitting reverence, & without fear of ridicule or ostentation of piety."[33]

Darwin spoke highly of the missionaries; he specifically mentioned Mr. Nott who had worked faithfully among the Tahitians for over forty years.[34] Nott had also won the praise and admiration of none other than Otto von Kotzebue who commended this missionary for his translation of the Bible, a Prayer Book, and some hymns into their language. Furthermore, "he also first instructed the Tahitians in reading and writing...."[35]

Darwin also credited Christianity with the abolition of drunkenness, human sacrifices, infanticide and the brutality of tribal wars that had occurred so regularly in the past. But even though he praised the ethical and educational efforts of the missionaries, he never once mentioned their work of evangelization among these people of the Pacific.[36]

After arriving in New Zealand, Darwin was disappointed with the impact that Christianity had upon the inhabitants there as compared to its reception in Tahiti. Nevertheless he recognized the superiority of the Christian faith over the heathen practices. Out of respect and admiration for the missionaries' efforts, "Darwin contributed to the large fund that *Beagle* officers raised for constructing a church in New Zealand."[37]

---

32 *Correspondence*, 1:472.

33 *Diary*, 1:320.

34 *Diary*, 1:322.

35 von Kotzebue, *A New Voyage Round the World*, 2:152.

36 Frank Burch Brown, *The Evolution of Darwin's Religious Views* (Macon, Georgia: Mercer University Press, 1986), 13.

37 Phipps, *Darwin's Religious Odyssey*, 27.

Like most Europeans, Darwin was baffled in Australia when he came upon such animals as the duck-billed platypus—a bizarre, egg-laying, venomous, beaver-tailed mammal! How does one account for such peculiar creatures? Darwin mused that possibly the Creator had made two distinct creations over a long period of time.[38] Even though he was by now a confirmed Lyellan uniformitarianist, Darwin still maintained that a Creator was absolutely necessary and was involved in his creation.

For all the success accomplished by the Christian missionaries in the Pacific, Darwin was convinced that they could never have any impact whatsoever upon the natives of Tierra del Fuego. The Fuegians, living at the tip of South America, "were probably the very lowest of the human race."[39] In a letter to Henslow dated April 11, 1833, after his first encounter with these people, Darwin wrote: "The Fuegians are in a more miserable state of barbarism, than I had expected to have seen in a human being."[40] A little over a month later, on May 23, 1832, he wrote to his cousin, William Darwin Fox: "I saw bona fide savages and they are as savage as the most curious person would desire. A wild man is indeed a miserable animal but one well worth seeing."[41]

It was not until 1867 that Darwin, upon reading about the spectacular results that the missionaries were having in Tierra del Fuego, wrote to Admiral Sir James Sulivan (1810-1890), a member of the South American Missionary Society, to tell him that he had been wrong. This James Sulivan was the same person who, as a lieutenant on board the *Beagle*, believed that no person "existed too low to comprehend the simple message of the Gospel of Christ."[42] Darwin, then fifty-eight, was so impressed with these Christian missionaries that he sent a cheque "as a testimony of the

---

38 *Diary*, 1:348.
39 *Life and Letters*, 2:308.
40 *Correspondence*, 1:306.
41 *Correspondence*, 1:316.
42 *Life and Letters*, 2:308.

interest he took in their good work"[43] and faithfully read the mission's newsletter to be informed of their activities and progress until his death in 1882.

## SEEING THE WORLD THROUGH LYELL'S EYES

If Darwin's views of God and the Bible had changed little, what did? From Darwin's perspective, the five-year trip was essentially an extended geological excursion. Through firsthand experience, he became convinced that the age of the earth spanned eons of time. What caused him to accept this uniformitarian view?

Having had little formal training in natural sciences and geology, Darwin reasoned: "Geology was a capital science to begin, as it requires nothing but a little reading, thinking and hammering."[44]

Scholars, after examining the copious notes made by this inexperienced naturalist during his fifty-seven month voyage, have concluded that geology was indeed his main preoccupation. Of the 2,530 pages of notes that he took, it should be no surprise that the breakdown would be as follows:

| | |
|---|---|
| Geological notes | 1,383 pages |
| Zoologial notes | 368 pages |
| Personal notes | 779 pages [45] |

Prior to leaving England in the fall of 1831, Captain FitzRoy had purchased Lyell's first volume of *Principles of Geology* for Darwin. Henslow had recommended it but, because of its denial of successive catastrophes in the formation of the earth's surface, "on no account was he to accept the views therein advocated."[46] Through the constant reading and re-reading of this first volume

---

43 *Life and Letters*, 2:308.

44 *Life and Letters*, 1:234f.

45 H. Gruber and V. Gruber, "The eye of reason: Darwin's development during the *Beagle* voyage," *ISIS* 53 (1962), 189.

46 *Autobiography*, 101.

and the two subsequent ones that were sent to him, Darwin became a confirmed uniformitarian. Consequently, he "became a zealous disciple of Mr. Lyell's views, as known in his admirable books."[47]

Lyell's epigrammatic principle, 'The present is the key to the past,' became Darwin's. As he studied the rock strata throughout the world, he became convinced that the geological strata had been laid down over millions of years and that natural processes had always been at work in altering it. Looking through Lyell's eyes, Darwin could see no evidence of a global catastrophe as outlined in the biblical narrative in Genesis.

From his *Beagle* days, Charles Darwin had nothing but praise for Lyell's work. On one occasion, in writing to Leonard Horner (1785-1864), Lyell's father-in-law, he said:

> I always feel as if my books came half out of Lyell's brain, and that I never acknowledge this sufficiently... for I have always thought the great merit of the *Principles* was that it altered the whole tone of one's mind.[48]

The *Principles of Geology* did indeed have a lasting impact on Darwin's thinking. Lyell, a Deist, had rejected the Bible and saw no need for God's intervention in the unfolding of geological history. Lyell's thoughts both geologically and religiously weighed heavy upon Darwin's mind. Lyell "who spoke to Darwin through the pages of the book became a model"[49] for what he was to become.

In September 1835, Darwin arrived at the Galapagos Islands. Being thoroughly persuaded by Lyell's teaching, he began for the first time to direct his attentions to the living creatures, particularly the finches on the island. How they differed from those on the mainland of South America! As one reared in a society committed to Natural Theology, he had always believed that animals were

---

47 *Life and Letters*, 1:234.
48 *More Letters*, 2:117.
49 Janet Browne, *Charles Darwin: Voyaging* (New York: Knopf, 1995), 324f.

created by God according to their kind and these kinds were immutable or fixed.[50] But could it be that these birds, separated from the mainland and subjected to environmental pressures over long periods of time as supplied by Lyell's theory, had changed? Were species, then, mutable? The question needed more time to ponder.

The *Beagle* docked back in England on October 2, 1836. When the twenty-seven-year-old world traveller returned home after a five-year absence, did his family notice any change in his religious views? The answer is no. His Wedgwood Unitarian heritage with its belief in a Creator who played an important role within his universe and its view of biblical morality was still, at least for the moment, intact. No clearer theological declaration can be found than in these words written near the end of his *Beagle* diary:

> Amongst the scenes which are deeply impressed on my mind, none exceed in sublimity the [Brazilian] primeval forests… or those of Tierra del Fuego, where death and decay prevail. Both are temples filled with the productions of the *God of Nature*. No one can stand unmoved in these solitudes, without feeling that there is more in man than the mere breath of his body.[51]

But one thing definitely *had* changed: Charles had abandoned forever any notion of a career in the Anglican Church.

In Darwin's absence, Henslow, his faithful mentor, had read his letters to the Philosophical Society of Cambridge.[52] This had already begun to open academic doors for his unique student. Armed with his voluminous notes, the large array of specimens he had sent to Henslow and with Lyell's principles of uniformitarianism as his

---

50 *More Letters*, 1:367.
51 *Diary*, 1:388. Italics are mine.
52 *Autobiography*, 81–82.

guide, Charles now had "the kind of geology he needed to become Charles Darwin, the evolutionist."[53]

---

53 Howard E. Gruber, *Darwin on Man: A Psychological Study of Scientific Creativity* (New York: E. Dutton, 1974), 90.

## Chapter 4
# Darwin's conversion to evolutionism

The year 1837 can be regarded as a significant date in British history as it marked the beginning of the sixty-three-year reign of Queen Victoria (1819-1901)—the longest reigning monarch in British history. This date was also important in the life of Charles Darwin because it marked his conversion to evolutionism.[1] If, according to Frank Sulloway,[2] March 1837 was the actual month, everything was in place for Darwin to give serious consideration to the species question.

Having settled in London where he could concentrate more fully on his *Beagle* diary, Darwin had an opportunity to meet with

---

1   Naturalism, like Supernaturalism, is a worldview and thus it answers the three eternal questions of life (see Preface). Naturalism in answering the question 'Where did we come from?' uses evolutionism, whereas Supernaturalism uses creationism.

2   "Darwin's Conversion: The *Beagle* Voyage and its Aftermath" in *Journal of the History of Biology* 15 (1982), 363f. The actual moment that Darwin became an evolutionist has sparked much debate among scholars, but there seems to be a consensus that it happened between 1837 and 1839.

John Gould (1804-1881), an eminent ornithologist and taxidermist. As a member of the Zoological Society of London, Gould had been studying Darwin's collection of birds and animals that had been donated to this prestigious society.

Prior to meeting Gould, Darwin had finished reading Lyell's fifth edition of the *Principles of Geology*. With the concept of uniformitarianism still fresh in his mind and with the suggestion from Gould that a large number of species from the Galapagos were new and found nowhere else, the penny dropped! Species, Darwin now reasoned, given long periods of time could change into new ones. Thus, "Darwin's scientific contacts with Gould at the time, appear to have been the final catalyst in his conversion to the theory of evolution."[3]

The fixity of kinds or the biblical belief that dogs produce only dogs, and cats produce only cats, as clearly taught in the Bible[4]—a belief generally accepted by everyone in Britain at that time—was considered by Darwin to be totally spurious. Not to be left in a religious vacuum, Charles Darwin, the evolutionist, was now wholeheartedly converted to Naturalism. In other words, he conceived a world controlled by natural laws devoid of any Divine interference.

Had not the Galapagos specimens, which had shown some evidence of transformation, especially the finches' beaks, been the key? Now, as one committed to evolutionism, when Darwin peered into the small world of the Galapagos, he perceived it as being, in miniature, a replica of what had been occurring universally over eons of time. The species barrier had been broken and the Bible could be cast aside!

Charles Darwin rejected the fixity of kinds and, for the same reason, denied a global flood—because of lack of faith in the biblical record. As we will show, it had nothing whatsoever to do with science, but rather was a conscious decision to accept another

---

3   Frank Sulloway, "Darwin's Conversion: The *Beagle* Voyage and its Aftermath," 369.

4   In the first chapter of Genesis, the word 'kind' is used ten times.

**CHARLES DARWIN (1809-1882)** • In his early thirties, Charles Darwin conceived the idea of evolutionism through natural selection.

belief system. Science, by definition, has always dealt with that which is observable and repeatable. The origins of kinds and the biblical flood all happened once, long ago in the past, and thus are beyond observation and outside the field of science.

"The existence of an Intelligent Creator was not in dispute; rather it was the Creator's chosen method of generating species on earth that was the central point of contention for Darwin."[5] Now, having committed himself to a belief that species were mutable or could be modified over long periods of time, Darwin pondered: "What was the naturalistic mechanism reponsible for changing one kind of life into another?"

## IN SEARCH OF A NATURALISTIC MECHANISM

Committed to the two assumptions that, first, the earth was eons of years old and that, second, modification (as witnessed among the finches of the Galapagos Islands) had occurred in living organisms, Darwin began his search to find this elusive mechanism. As had been his custom on the *Beagle*, he began jotting down ideas in small notebooks. In March, he opened his "Red Notebook," in July[6] of that year, he subdivided it into books: "A" for geology (little of this book has survived) and "B" for his evolutionary ideas.

Across the top of the first page of "B", he wrote "ZOONOMIA"—the name that his grandfather, Erasmus, had titled his book. The younger Darwin had just re-read his illustrious grandfather's book but was disappointed that his treatment of evolutionary development was much too speculative to convince the scientific community of the day. He would produce a "NEW ZOONOMIA"—a naturalistic approach in unravelling the species question.

From July 1837 to July 1839, Darwin filled six notebooks with casual observations, flashes of insights or meaningful ideas from

5   Michael A. Corey, *Back to Darwin: The Scientific Case for Deistic Evolution* (Lanham: University Press of America, 1994), 8.

6   *Autobiography*, 83.

> **B, C, D, E** ———————→ *On the Origin of Species* (1859)
>
> **M, N** ———————————→ *The Descent of Man* (1871)

**THE SOURCE FOR DARWIN'S BOOKS** • The chart above shows which of his notebook subdivisions formed the basis of his major works.

his readings or private conversations, especially with his father.[7] "Entries in the notebooks were short, careless of traditional elements of style and frequently obscure."[8] But even within these limitations, they were a man's private musings and were an accurate portrayal of his intellectual development. Undoubtedly, these notebooks were the creative reservoir from which all of Darwin's subsequent works would flow, as illustrated above.

By October 1, 1838, some fifteen months after opening his first notebook, it all seemed so simple. Five days before, Charles had begun reading *Essays on the Principle of Population* (1798) by Thomas Malthus (1766-1834) and he "had at last got a theory by which to work."[9]

Malthus' basic theory was that there was an ongoing struggle between population growth and the availability of food. He pessimistically forecasted that the latter would in time be outstripped by the former. But, he also realized that famines, plagues, poverty and war played a significant role in keeping the increase of human population in check. "The insight Darwin found in Malthus' work supplied him with an essential element for completing his biological theory."[10]

---

7  Edward Manier, *The Young Darwin and His Cultural Circle: A study of influences which helped shape the language and logic of the first drafts of the theory of natural selection* (Reidel: Dordrecht, Holland, 1978), 18-19. A table shows 46 individuals who were mentioned in his notebooks and manuscripts. His father tops the list.

8  Sandra Herbert, "The Place of Man in the Development of Darwin's Theory of Transmutation, Part 2" in *The Journal of the History of Biology* 10 (1977), 180.

9  *Autobiography*, 120.

10  William E. Phipps, *Darwin's Religious Odyssey* (Harrisburg: Trinity Press International, 2002), 38.

**THOMAS MALTHUS (1766-1834)** • Malthus was a professor of history and economics. His theory of the relationship between food supply and population growth became the basis for Darwin's evolutionary theory.

The concept of an ever-present struggle for existence immediately captured Darwin's attention. Now, for the first time, he saw that all living things were engaged in a relentless battle with their environment to survive.

> It at once struck me that under these circumstances favourable variations would tend to be preserved, and unfavourable ones to be destroyed. The result of this would be the formation of new species. Here then I had at last got a theory by which to work; ...[11]

But what was the mechanism that caused the formation of these new species? He reasoned that natural selection in concert with environmental changes over long periods of time was able to generate and develop new life forms. In his mind, the enigma was resolved.

Now, having the framework of 'his theory,' Darwin began his research. He considered that there must be some connection between natural selection in nature and artificial selection by humans. So, in January 1839, he started to correspond with breeders of animals and horticulturalists.

As an evolutionist who believed in a naturalistic explanation for the development of life, the question that now had to be addressed was: How would Darwin's religious views be affected?

## THE REJECTION OF 'BIBLICAL' UNITARIANISM

Between 1836 and 1839, Darwin wrote in his *Autobiography*: "I was led to think much about religion."[12] As one committed to Naturalism, Darwin's theory attacked the Bible, the foundation of his Unitarianism:

> The Old Testament from its manifestly false history of the

---

11 *Autobiography*, 120.
12 *Autobiography*, 120.

world, with the Tower of Babel, the rainbow as a sign, etc.,
and from its attributing to God the feelings of a revengeful
tyrant, was no more to be trusted than the sacred books
of the Hindus, or the beliefs of any barbarian.[13]

As Frank Burch Brown reminds us, that last statement is most con-
demning "given Darwin's low opinion of barbarians."[14]

Darwin no longer believed the miraculous in the Old Testament
nor those in the New Testament. He reasoned:

…the more we know of fixed laws of nature the more
incredible do miracles become,—that the men at that time
were ignorant to us… that the Gospels cannot be proved to
have been written simultaneously with the events…[15]

Even with this highly derogatory assessment of the New Testament
authors, the former theological student commended their writings
for their teaching on morality. He lamented that the New Testament
did not instruct that the pursuit of happiness came primarily
through intellectual cultivation and, secondly, through helping
others. His skepticism concerning the Old Testament was heightened when he considered that Lyell's uniformitarian principles of
an earth spanning eons of time contradicted the then-accepted
biblical chronology of thousands of years.

Furthermore, Darwin's animosity towards 'biblical' Unitarianism evidenced itself concerning the question of divine justice. He
was aware that the natural reading of the biblical text stated: "The
Lord will judge his people. It is a fearful thing to fall into the hands
of the living God."[16] Divine judgement on unbelievers, he realized,

---

13 *Autobiography*, 85.
14 Frank Burch Brown, *The Evolution of Darwin's Religious Views* (Macon, Georgia:
Mercer University Press, 1986), 13.
15 *Autobiography*, 86.
16 Hebrews 10:30-31.

would include "his Father, brother and almost all his best friends [who would] be everlastingly punished. And this was a damnable doctrine."[17]

Darwin's rejection and repugnance of divine judgement was totally consistent with his newly-acquired belief in Naturalism. His lack of spiritual discernment was most evident here. Dismissing the Bible as the Word of God, Darwin had no basis upon which to believe that a holy, righteous God abhorred sin and would judge men and women for their rebellion. But Augustus Comte (1798-1857) would provide him with the perfect solution to resolve this dilemma of a righteous God and a sinful mankind.

On Lyell's insistence, Darwin was elected to the Athenæum Club on June 21, 1838 along with Charles Dickens (1812-1870).[18] This exclusive club allowed Charles to mingle with London's most influential thinkers. On August 12, while visiting the club, he read a review of *The Course of Positive Philosophy* by Comte, a French philosopher. As an atheist, this founder of modern sociology had developed a theory of a three-staged system that humans had gone through in order to reach their present level of understanding of how the world in which they lived operated. They were: theological, metaphysical and positive. The first and the most primitive stage placed mankind in a state of using the supernatural agencies to explain the natural phenomena around him. Secondly, as mankind became more knowledgeable and sophisticated, he totally renounced these religious attachments; in the third, or positive stage, he had the capability of understanding his environment by observing the lawful acts of nature. Everything, according to Comte's positivism, could be reduced to some naturalistic explanation, even to the development of religions themselves.

Darwin immediately saw the usefulness of Comte's position in eliminating the necessity of God from the operation of a lawful universe. He argued that the philosophers were wrong in saying

---

17 *Autobiography*, 87.
18 *Correspondence*, 2:94, footnote 2.

that people were born with an innate knowledge of a Supreme Being.[19] Could not mankind have conjured up the need for God, or gods, as they looked with fear and amazement at the majesty and awesomeness of nature? Indeed, they did! All religions, including Christianity with its righteous God and sinful mankind, were in reality a product of man's fertile imagination to explain and cope with the vagaries of nature. It was inevitable that, after reaching this conclusion about religions, Charles Darwin, a young man of thirty, "gradually came to disbelieve in Christianity as a divine revelation."[20]

Darwin had some difficulty in accepting Comte's atheism but he subscribed to the stages of mankind's religious development. The Bible was indeed an ancient book written when mankind lacked the sophistication and intellectual development that he had achieved by the late nineteenth century.

### EMMA DARWIN'S FAITH

To complicate matters for this young evolutionist, Charles turned his affections to Emma Wedgwood (1808–1896), his first cousin and the daughter of his favourite uncle, Josiah Wedgwood II.[21] For the time period, Emma, the youngest daughter of Josiah, was well educated. Her education included frequent trips to Europe where she became proficient in French, Italian and German. An accomplished pianist, she studied under Frédéric Chopin (1810–1849) in France.

After a short engagement, Charles and Emma were married on January 19, 1839, at St. Peter's Anglican Church, Maer, Staffordshire, by their mutual first cousin, Rev. John Allen Wedgwood (1796–1882). Charles' and Emma's first residence was a home they rented in London. There they attended the Unitarian Chapel on Little Portland Street, worshipping with Hensleigh, Emma's brother, and his wife, Fanny.[22]

---

19 Romans 1:19-20.
20 *Autobiography*, 86.
21 See the Wedgwood-Darwin family tree on page 3.
22 Randal Keynes, *Annie's Box: Charles Darwin, His Daughter and Human Evolution*

**EMMA WEDGWOOD DARWIN (1808-1896)** • Emma married her first cousin, Charles Darwin, in 1839 at age 31, the same year she sat for a portrait by George Romney (this is a sketch after that portrait).

At the age of sixteen on September 7, 1828, Emma had been confirmed in the same Anglican church in which she was married. As a typical Wedgwood, she was "Unitarian by conviction, Anglican by practice."[23] Her mother, Elizabeth Allen (1764-1846), having a strong social conscience, conducted a Sunday School for the poor in the servants' quarters. Not only were they taught the 'three Rs' but they were exposed to Unitarian teachings about the Bible.[24] This tradition was carried on by Emma and her sister, Elizabeth; they held the classes in the laundry room of their home and had as many as sixty youngsters attending.

Charles' and Emma's daughter, Henrietta Litchfield (1843-1929), had these recollections of her mother:

> She went regularly to church and took the Sacrament. She read the Bible with us and taught us a simple Unitarian Creed, though we were baptized and confirmed in the Church of England.[25]

To the socially conscious, allegiance to the Anglican Church in eighteenth-century Britain was never questioned. Both Charles and Emma were very sensitive to this unwritten expectation; thus, they saw the necessity to be involved in the local Anglican church within the Downe community, even though neither shared its beliefs. At least for the Darwins, social respectability trumped theology.

James Moore states: "Emma was a sincere believer in the Christian plan of salvation and those who trusted in Jesus and his resurrection from the dead would spend eternity in Heaven."[26] No clearer decla-

---

(New York: Riverhead Books, 2002), 10.

23 James Moore, *The Darwin Legend* (Grand Rapids: Baker Books, 1994), 36.

24 B. and H. Wedgwood, *The Wedgwood Circle, 1730-1897* (Westfield, N.J.: Eastview, 1980), 195.

25 Henrietta Litchfield, ed., *Emma Darwin: A Century of Family Letters, 1792-1896*, 2 vols. (London: John Murray, 1915), 2:173.

26 *Evolution: Darwin's Dangerous Idea*, written and directed by David Espar. 120 min. Clear Blue Sky Production, 2001, videorecording (DVD).

ration of New Testament truth could be stated. Consequently, many scholars are convinced that Emma was an evangelical Christian.[27] But this question must be asked: Who was this Jesus in whom she had placed her faith?

As a 'biblical Unitarian,' Emma believed in one eternal God; Jesus was a creation of God. Living a perfect life, this same Jesus provided eternal life through his death, burial and resurrection for all those who were morally good. During the morning church service, the Darwins regularly showed their disdain for the trinitarian doctrine so clearly taught in the Anglican *Thirty-Nine Articles.* "When the congregation turned towards the altar to recite the Creed, the Darwins faced the other way and sternly looked into the eyes of the other church-goers."[28]

It should be noted that Emma's view of the Scriptures was anything but orthodox. When reading Charles' *Autobiography* after his death, she penned a note questioning the doctrine of verbal inspiration beside a passage referring to everlasting punishment. Prior to its publication, she had this passage deleted.[29] No one should doubt Emma's reverence for the Bible or her frequent reading of it. But her anxiety over her husband's renunciation of the Bible became well known through her extant letters.

Prior to their engagement, Emma was aware of the doubts about her 'biblical' Unitarianism brewing within the mind of her husband-to-be. Charles, contrary to his father's advice, had been quite open and honest with her concerning his uncertainties regarding the Bible. But his innermost thoughts, as recorded in his notebooks,

---

27 "Emma's Christianity was a simple evangelical prescription…" [A. Desmond and J. Moore, *Darwin: The Life of a Tormented Evolutionist* (New York: Warner, 1991), 281]; "Emma, who was always a steadfast Victorian Christian…" [Phipps, *Darwin's Religious Odyssey,* 154]; and, "Emma wanted Charles Darwin to understand how important her Christian faith was" [Edna Healey, *Wives of Fame* (London: Sedgwick & Jackson, 1986), 156].

28 Keynes, *Annie's Box,* 128.

29 Some 6,000 words were expurgated prior to the publication of her husband's *Autobiography.* The vast majority were his criticisms of Christianity.

were not revealed to her.

On November 23, 1838, a week or so after their engagement, Emma wrote to Charles asking him to do something for her:

> Will you do me a favor? Yes, I am sure you will; it is to read our Savior's farewell discourse to his disciples at the end of the 13th chapter of John. It is so full of love to them and devotion and every beautiful feeling. It is part of the New Testament I love best.[30]

Whether Darwin read this portion of John's gospel and what his reaction was is unknown. But Emma's concerns over her husband's disbelief persisted. Two letters, one written shortly after their wedding and the other some twenty years later (1861), reveal an abiding anxiety. She realized that "Darwin's habitually scientific way of thinking, along with the anti-religious influence of his brother Erasmus had made him mistrust truths that lie beyond the bounds of science."[31] She cautioned him:

> I should say also there is a danger in giving up revelation which does not exist on the other side, that is the fear of ingratitude in casting off what has been done for your benefit as well as for that of all the world and which ought to make you still more careful, perhaps even fearful lest you should not have taken all pains to judge truly.[32]

The deistic influence of Lyell, Charles' older mentor and close friend, had undoubtedly taken its toll. Deism with its appeal to reason as the sole arbiter of truth regarded any type of divine revelation as suspect. Emma believed the greatest peril that her beloved faced in rejecting the Bible was that he would not participate in

---

30 *Correspondence*, 2:123.
31 Burch Brown, *The Evolution of Darwin's Religious Views*, 35.
32 "Mrs. Darwin's Papers on Religion" in *Autobiography*, 236.

the resurrection from the dead with her. This was a hope that she cherished. With utmost sincerity, she wrote: "Every thing that concerns you concerns me & I should be most unhappy if I thought we did not belong to each other forever."[33]

From the postscripts penned by Charles on both letters, it is evident that Emma's loving concern deeply touched her husband and her letters weighed heavily on his soul. On the first letter, he wrote: "When I am dead know that many times, I have kissed and cried over this."[34]

From his notebooks of the period (1837 to 1839), Darwin made it plain that, in the adherence to Naturalism, there was no need for God to be involved in the ongoing processes of the world since the beginning. Natural laws in every sphere, whether they be in the physical or moral, were entirely sufficient to account for change and development.

Darwin's ambivalence concerning God's existence has given scholars a great deal of latitude in pinning down his religious persuasion during this most critical period of his life—they span from his being a theist (more correctly, a deist), to agnostic, or even an atheist.[35]

---

33 *Correspondence*, 2:172.

34 "Mrs. Darwin's Papers on Religion" in *Autobiography*, 237.

35 For an excellent discussion on the position of Darwinian scholars, see Burch Brown, *The Evolution of Darwin's Religious Views*, 17-20.

## Chapter 5
# Divergent spiritual journeys

During the nearly six years that Charles Darwin and Captain FitzRoy spent together on the *Beagle* voyage, a bond of mutual respect developed. They had much in common: a fascination for the wonders of geology and a similar religious outlook. But, within a period of two years after returning to England, these two men would take divergent religious paths. By moving into two different spheres of belief, what impact, if any, was there on their relationship?

### FROM CADET TO CAPTAIN OF HMS *BEAGLE*

Robert FitzRoy was a true aristocrat. "The FitzRoys had descended as the dukes of Grafton, a favoured though illegitimate branch of royalty, from a liaison between Barbara Villiers and King Charles II."[1] Three of this lineage became British admirals and Robert's father,

---

1  Peter Nichols, *Evolution's Captain* (New York: Harper Collins, 2003), 21.

Lord Charles FitzRoy (1764-1829), attained the rank of general in the British army. Robert's mother, Lady Frances Anne Stewart (d.1810), bestowed on her son an Irish heritage as she was the eldest daughter of Robert Stewart, 1st Marquis of Londonderry (1739-1821).

After attending two grammar schools, Robert FitzRoy entered the Portsmouth Naval College at the age of thirteen. Some twenty months later, he graduated with distinction, attaining a gold medal in mathematics. In 1819, he went to sea and served in the Mediterranean and South American fleets. Over a nine-year period, he passed "through the ranks of Volunteer-per-order, College Volunteer, Midshipman, Lieutenant and Flag Lieutenant."[2]

Sir Robert Otway (1770-1846), the commander-in-chief of the South American fleet and captain of HMS *Ganges*, was impressed with the naval ability of Robert FitzRoy, his Flag Lieutenant. So, when the captaincy of HMS *Beagle* became vacant with the suicide of Captain Pringles Stokes (d.1828), Admiral Otway commissioned Lieutenant FitzRoy to replace him. FitzRoy took command in Rio de Janeiro on December 15, 1828. The appointment by Otway overruled the obvious choice of Lieutenant William Skyring (d.1833), the acting commander after Stokes' death.

Over the next two years, the twenty-three-year old captain of HMS *Beagle* completed the surveying tour of duty. But on his return to England on September 12, 1830, and much to the consternation of the British Admiralty, Captain FitzRoy informed his commanding officer, Captain Phillip Packer King, that he had four captured natives of Tierra del Fuego on board. Their names and estimated ages were: York Minster, 26; Boat Memory, 20; Jemmy Button, 14; and Fuegia Basket (a girl), 9. Furthermore, he added: "I have maintained them entirely at my own expense, and hold myself responsible for their comfort while away from, and for their safe return

---

2 H.E.L. Mellersh, *FitzRoy of the Beagle* (London: Rupert Hart-Davis, 1968), 21.

to their own country."[3] His goal was to expose them "to the plainer truths of Christianity, as the first object; and the use of common tools, a slight acquaintance with husbandry, gardening, and mechanism, as the second."[4]

During the year that the three (Boat Memory had died from a small pox vaccination) were in England, they stayed at St. Mary's Infants' School at Walthamstow under the care of Rev. William Wilson and his wife. Here, they were exposed to the basic truths of Christianity. As expected, people's curiosity to see them gave ample opportunity to showcase these three Fuegians throughout England. In the late summer of 1831, they, along with Captain FitzRoy, were ushered into the presence of none other than King William IV and Queen Adelaide of England.

Much to his own relief, FitzRoy's plan to keep the Fuegians in England for two or three years never came to fruition. On July 4, 1831, the British Admiralty commissioned HMS *Beagle* into action with FitzRoy at its helm. His main task was to complete the surveying of both sides of the southern coasts of South America. It was also recommended that two Anglican missionaries be sent to minister to the people of Tierra del Fuego. In the early decades of the nineteenth century, Anglican missions were viewed as a means of transposing British culture and values into a foreign land. Richard Matthews (1811-1893), a totally inexperienced twenty-year-old who had a brother serving in New Zealand, was the only one to volunteer. His mandate was to instruct "a core of Christianity by showing the native Fuegians how to farm the land and build wooden houses, giving the people clothes, promoting cleanliness, and, if all went well, teaching some basic English and the precepts of the established Anglican Church."[5] Even with their brief stay in

---

3   Robert FitzRoy, *Narrative of the Surveying Voyages of His Majesty's ships Adventure and Beagle between 1826-1836* (1839), 4. Hereafter, it will be cited as *Narrative*.

4   *Narrative*, 12.

5   Janet Browne, *Charles Darwin: Voyaging* (New York: Knopf, 1995), 239.

**VICE-ADMIRAL ROBERT FITZROY (1805-1865)** • FitzRoy was captain of HMS *Beagle* when it made its surveying voyages to the South Pacific. He was a pioneer in meteorology and weather forecasting, an accomplished surveyor and hydrographer. FitzRoy also served as the second governor of New Zealand (1843–1845).

England, it was believed that the three returning Fuegians would be invaluable in the Anglicanization process.

Immediately, the *Beagle* was sent to the dock to be refitted. Captain FitzRoy personally supervised the refit. "No expense seemed to have been spared in preparing her for the voyage, with a considerable amount of that expense coming from FitzRoy's pocket, for items that he considered essential but the Admiralty regarded as luxuries."[6] Structurally, he had the main deck raised to 1.8 m (6') to increase the headroom on the lower deck.

Lightning storms were a ubiquitous danger for seafarers. To provide safety for both his ship and crew, FitzRoy, who kept himself abreast of the newest scientific advancements, installed lightning conductors. Invented by Sir William Snow Harris (1791–1867), they were continuous strips of copper tape from the mast to the keel and thus to the water.[7] During the nearly five-year trip, the ship was struck by lightning but never experienced any damage. FitzRoy was definitely ahead of his time. It was not until 1847, some sixteen years later, that the British Government would mandate this extremely important safety device for all seafaring vessels.

## LIFE TOGETHER ON THE *BEAGLE*

As noted in chapter 3, Captain FitzRoy, not wanting to endure the isolation of being the ship's captain, asked his superior officer, the renowned explorer and hydrographer Sir Francis Beaufort (1774–1857), if he could be granted permission to have a mess companion. This person was to be "an intellectual and social equal, whose company would act as a safety valve"[8] to relieve the loneliness

---

6 John and Mary Gribbon, *FitzRoy: The Remarkable Story of Darwin's Captain and the Invention of the Weather Forecast* (New Haven: Yale University Press, 2004), 82.

7 "Appendix E" in Keith S. Thomson, HMS *Beagle: The Story of Darwin's Ship* (New York: Norton, 1995), 294–296.

8 Gribbon, *FitzRoy: The Remarkable Story of Darwin's Captain*, 86.

that often beset those in command. It was at this juncture in his life that Charles Darwin stepped onto the stage.

These two young men, both in their twenties, spent many amicable hours together during their circumnavigation of the world. In a letter to his sister, Caroline, on April 25, 1835, Charles wrote this glowing report of his captain:

> As far as I can judge: he is a very extraordinary person.—I never before came across a man whom I could **fancy** being a Napoleon or a Nelson.—I should not call him clever, yet I feel convinced nothing is too great or too high for him.— His ascendency over every-body is quite curious:… he is the strongest marked character I ever fell in with.[9]

During the five-year voyage, FitzRoy, an Anglican by birth only, rarely came into conflict with Darwin on religious matters. In reality, their belief systems were almost identical. Following the custom of the day, FitzRoy's parents ensured that he, like his mess confrere, participated in the rites of passage afforded by the Anglican Church. As the captain of the *Beagle*, he was required to conduct daily regular Church of England services, which included Scripture reading. Darwin also routinely attended.

That God was the Creator and that all mankind were related through Adam and Eve, as recorded in the Bible, was accepted as an undeniable fact. Even the most primitive was part of the same human stock. FitzRoy was convinced that, if these people were given the proper training, they could become as educated and cultured as any European. He personally witnessed the transformation of three Fuegians "into well behaved, civilized people, who were very much liked by their English friends."[10] FitzRoy saw

---

9   *Correspondence*, 1:226-227. The word 'fancy' was bold in the original. See also *Autobiography*, 72-73.

10   Robert FitzRoy and Charles Darwin, "A Letter, Containing Remarks on the Moral State of Tahiti, New Zealand, &c" in *South African Christian Recorder* 2, No. 4

Boat Memory's premature death as a personal tragedy. In his inter-action with his four Fuegian 'captives,' FitzRoy sensed that Boat Memory was by far the most intelligent and showed the greatest promise in becoming totally Anglicized.[11]

In a letter to his sister Caroline in 1833, after having come in contact with Fuegians for the first time, Darwin made this remark-able comment—especially in light of his position some forty years later in his *Descent of Man*:

> We here saw the native Fuegian; an untamed savage is I really think one of the most extraordinary spectacles in the world… in the naked barbarian, with his body coated with paint, whose very gestures, whether they may be peaceable or hostile are unintelligible, with difficulty *we see a fellow-creature.*[12]

Their lifestyle was definitely considered to be inferior to those in Britain but they were "just as much brothers under the skin."[13]

With his scientific bent, FitzRoy was very much interested in the newest branch of science—geology. Like Darwin, his fellow naturalist, he was very dependent upon Lyell's three volumes of *Principles of Geology* in understanding the geological formations throughout the trip. Foundational to any Lyellian interpretation was the denial of a global flood as outlined in Genesis. Cambridge scholars like John Henslow and Adam Sedgwick had publicly questioned the authenticity of the biblical account. Later, on his return to England, FitzRoy demonstrated a degree of honesty and insight by admitting:

> While led away by sceptical ideas, and knowing extremely

---

(September, 1836), 222.
11 *Narrative*, 1:11.
12 *Correspondence*, 1:302-303. Italics not in the original.
13 Browne, *Charles Darwin: Voyaging*, 245.

little of the Bible, one of my remarks to a friend, on cross-
ing vast plains composed of rolled stones bedded in diluvial
detritus some hundred feet in depth, was "this could never
have been effected by a forty days' flood,"—an expression
plainly indicative of the turn of mind, and ignorance of
Scripture. I was quite willing to disbelieve what I thought
to be the Mosaic account, upon the evidence of a hasty
glance, though knowing next to nothing of the record I
doubted.[14]

Only on one occasion when they were discussing the question
of slavery did a serious disagreement erupt. In his *Autobiography*,
Darwin described the quarrel. After visiting a coffee plantation,
FitzRoy was convinced that the slaves there were happy. In a conversation with them, he learned that they also stated that they had
no desire to be emancipated. Such gullibility angered Charles.
How else, he retorted, would you expect these people to respond
in the presence of their master! "This made him excessively angry,
and he said that as I doubted his word, we could not live together."[15]
Later, FitzRoy apologized and Darwin rejoined him in his cabin.
By mutual agreement, the topic was never discussed again.

Both the Darwins and the Wedgwoods were known for their
condemnation of slavery. Reared within this atmosphere, Charles
too had come to detest its degradation and dehumanization. Josiah
Wedgwood I, his maternal grandfather, had won the acclaim of the
Society for the Abolition of the Slave Trade for his productions of
"hundreds of medallions bearing the image of a chained African
saying, 'Am I not a man and a brother?'"[16]

When Charles was studying at the University of Edinburgh, he
met John Edmonstone, a freed South American slave, who did

---

14 "A Few Remarks with Reference to the Deluge" in *Narrative*, 658-659.

15 *Autobiography*, 74.

16 William E. Phipps, *Darwin's Religious Odyssey* (Harrisburg: Trinity Press International, 2002), 22.

**ANTI-SLAVERY MEDALLIONS** • Josiah Wedgwood I, Darwin's grandfather, had produced hundreds of anti-slavery medallions at his Staffordshire potteries.

taxidermy on a freelance basis for the university. Darwin hired this highly competent taxidermist to teach him this skill. They spent some time together and Darwin regarded him as "a very pleasant and intelligent man."[17]

In her excellent biography, Janet Browne correctly pointed out Darwin's hypocrisy. Back in England, both his grandfathers amassed great fortunes "on the backs of entrepreneurial companies that exploited cheap labour"[18] in the building of canals, railways or the thousands working at the Wedgwood pottery factory. This human bondage, not addressed until the labour legislation of the 1840s, was definitely a blind spot in Darwin's social conscience.

---

17 *Autobiography*, 51.
18 Browne, *Charles Darwin: Voyaging*, 245.

Never an advocate of slavery, Captain FitzRoy had developed a type of paternalism towards indigenous people. This attitude is best reflected in a letter that he and Darwin wrote to the *South African Christian Recorder* while at sea, on June 26, 1836. While staying in Cape Town, South Africa, they had encountered some hostility toward Christian missionaries. Feeling that the criticism was unjust and having had such a positive experience with the missionaries in Tahiti and New Zealand, they felt compelled to respond in their defence. In a seventeen-page letter, including excerpts from their respective diaries, they outlined the beneficial effects of Christian-ization upon the native populace. Comments ranging from dress, manners, morals and education were used to convince the most ardent South African sceptic that the missionaries not only modelled the best of Western values but they instilled Christian care and compassion through the example of their daily lives. Their letter concluded with this admonition:

> On the whole, balancing all that we have heard, and all that we ourselves have seen concerning the missionaries in the Pacific, we are very much satisfied that they thoroughly deserve the warmest support not only of individuals, but the British Government.[19]

Their collaborative effort clearly demonstrated the bond of friend-ship between these two shipmates; nothing divided them religiously. At this point in their lives, there seemed little that could jeopardize their friendship. On October 2, 1836, some fourteen weeks later, the *Beagle* docked at Falmouth, England. The relationship between the captain and his messmate—which had extended for nearly five years—had formally come to an end. The ensuing years were to lead them in spiritual paths that were diametrically opposite.

On his return to England, Captain FitzRoy was honoured

---

19 FitzRoy and Darwin, "A Letter, Containing Remarks on the Moral State of Tahiti, New Zealand, &c," 238.

publicly for his noteworthy accomplishments. Recognition for his work was acknowledged by the British Parliament. "He had proved himself an extremely competent marine surveyor and a producer of charts that are remembered as examples of their kind in the Royal Navy to this day."[20] In 1839, he received the Gold Medal from the Royal Geographical Society.

## ROBERT FITZROY'S CONVERSION TO CHRISTIANITY

> Much of my own uneasiness was caused by reading works written by men of Voltaire's school; and by those geologists who contradict, by implication, if not in plain terms, the authenticity if the Scriptures; ...For men who, like myself formerly, are willingly ignorant of the Bible, and doubt its divine inspiration, I can only have one feeling—sincere sorrow.[21]

This heart-felt confession was written by FitzRoy in the final chapter of his *Narrative* published in 1839. While circumnavigating the world, both FitzRoy and Darwin agreed that they would publish together the accounts of their *Beagle* experience—but in separate volumes. When Darwin received FitzRoy's, he was aghast when he read the last two chapters: "Remarks on the Early Migration of the Human Race" and "A Few Remarks with Reference to the Deluge."

Darwin was very much aware that Captain FitzRoy had experienced a spiritual awakening some time after their arrival home. But to publish material that accepted the biblical account of the origin and migration of mankind and also the historicity of Noah's flood was, from Darwin's perspective, outlandish. To his sister Caroline, he wrote: "You will be amused with FitzRoy's Deluge

---

20 Mellersh, *FitzRoy of the Beagle*, 172.
21 "A Few Remarks with Reference to the Deluge" in *Narrative*, 658-659.

Chapter—Lyell, who was here today, had just read it, & he says it beats all other nonsense he has ever read on the subject."[22]

That Lyell was miffed with FitzRoy's remarks is understandable. He realized that the chapter on the Noahic deluge was a direct attack on uniformitarianism—the underpinning of his *Principles of Geology.* Moreover he, along with other British geologists, was accused of leading FitzRoy astray from biblical truth. To warn young sailors not to fall into the same pit of disbelief and despair that he had, now became FitzRoy's driving passion. Even though these chapters were highly controversial, it was of paramount importance that biblical truth be sounded forth for young minds to grasp.

What is the basis for believing that FitzRoy had a born-again[23] experience? To date, no record, either personal or otherwise, has been uncovered that actually describes his conversion to Christianity, but there are a number of very compelling inferences that point in that direction.

On December 8, 1836, he married Maria Henrietta O'Brien (1812-1852), daughter of Major-General Edward O'Brien. The ceremony took place almost two months to the day from the *Beagle's* arrival back in England; the announcement took Charles completely by surprise. He referred to Maria as "very beautiful and religious."[24] Being evangelical, she accepted the Bible as the Word of the living God. Even though little is known about the religious relationship between FitzRoy and his wife, she undoubtedly had a profound effect on his new commitment to Christianity.

Since FitzRoy's conversion coincided with the writing of his *Narrative*, this book is an excellent source in gleaning his religious

---

22 *Correspondence*, 2:236.

23 In John 3:3, Jesus said, "Truly, truly, I say to you, unless one is born again [or from above], one is not able to see the kingdom of God." In other words, a person must receive from God a *new* nature or identity to replace the natural or Adamic one that despises God. See Psalm 58:3; Romans 3:10-18.

24 *Correspondence*, 2:236.

insights as they were developing within this time frame. His *Narrative* reveals the tremendous admiration that he had towards the missionaries both in Tahiti and New Zealand. Here he witnessed first-hand, men who had a living and vibrant biblical faith. Their love for the native people and the sincerity of their prayer life had a lasting impact upon FitzRoy.

As was Darwin, Captain FitzRoy was thoroughly impressed with the missionary Mr. Nott, both as a person and with regard to his work on translating the Bible. Two years after FitzRoy had initially met him in Tahiti, this veteran missionary visited England. He made a special effort to visit FitzRoy in order to present him with a personal copy of the Bible translated into the Tahitian language. There definitely seems to have developed a special spiritual bond between these two men. FitzRoy "felt deeply gratified by that good man's kindness in giving him one of the first copies which were printed."[25]

Both in Tahiti and in New Zealand, Captain FitzRoy mentioned the importance of the truths of the gospel—the good news of the salvation message of Jesus Christ—being preached.[26] Evangelization took precedence over bringing European values and culture to these indigenous people. There was also a deep sense of gratitude when this British mariner heard that the missionaries in New Zealand were concerned about the spiritual welfare of the Fuegians. This attitude, from FitzRoy's perspective, was at the very heart of Christian missions. Furthermore, these missionaries refused to acknowledge that Richard Matthews' attempt to reach the Fuegians was a failure. They admitted, "It was the first step and similar in its result to our first step in New Zealand. We failed at first; but by God's blessing upon human exertions, we have at last succeeded far beyond our anticipations."[27]

After the publication of their books, FitzRoy's and Darwin's friendship waned as a result of the different spiritual paths that they

25 *Narrative*, 517.

26 *Narrative*, 541, 604.

27 *Narrative*, 605.

chose to take. Nonetheless, Darwin never lost his respect and admiration for his former sea captain. In 1851, he even supported FitzRoys' election to the Royal Society of London.

In 1859, Charles sent FitzRoy, now Rear-Admiral and head of the Meteorological Department, a copy of *On the Origin of Species*.[28] The content of this book served to cause an even deeper rift between the two of them. FitzRoy felt compelled to publicly denounce the fallacies of evolutionism. When a letter written in *The Times* supporting this heresy sparked his ire, he responded under the *nom de plume* of Senex. Darwin knew at once that it was his former captain.

In 1860, a meeting for the British Association for the Advancement of Science was held at Oxford University. As head of the Meteorological Office, Robert FitzRoy made a presentation on Friday night in which "he outlined the forecasting powers of barometric readings and described how his Met office was engaged in receiving information and providing forecasts by telegraph."[29] But the main attraction was on Saturday: Dr. John William Draper (1811-1882), head of the medical school at City University in New York, spoke on the topic: "On the intellectual development of Europe considered with reference to the views of Mr. Darwin." Darwin was absent as he was ill.

Draper's presentation was followed by a spirited debate between Samuel Wilberforce (1805-1873), Bishop of Oxford, and Thomas Henry Huxley (1825-1895) on the merits of evolutionism. The *Athenaeum* was the only official written record of the proceedings and it never mentioned the verbal exchanges between these two.[30] It is unfortunate that later historians have relied too heavily on

---

28 *Correspondence*, 7:376-377.

29 Nichols, *Evolution's Captain*, 315.

30 "Report of the British meeting in Oxford, 26 June—3 July 1860" in *Correspendence*, 8:595.

the biased accounts by Huxley himself and his supporter, Joseph Dalton Hooker (1817-1911). There was one reference to FitzRoy's involvement. It stated: "Admiral FitzRoy regretted the publication of Mr. Darwin's book and denied Professor Huxley's statement that it was a logical arrangement of facts."[31]

It was also during this time in his life that FitzRoy was under a great deal of pressure as head of the Meteorological Office. The responsibility of making weather forecasts (a term he coined) that were reliable was a daunting task. It became more onerous for the former naval officer when many seafarers became dependent upon them. Unfortunately, the accuracy of his daily forecasts came under fire and FitzRoy personally took the brunt of the ridicule. Not heeding the medical advice that he was given to take a necessary reprieve from the 'Met,' he eventually took his own life on April 30, 1865.[32]

Many of FitzRoy's scientific achievements, especially in meteorology, are remembered to this day, but the Scripture passages on his tombstone continue to speak volumes of his enduring faith. It reads:

---

31 *Correspondence*, 8:595. Interestingly, Mellersh (Mellersh, *FitzRoy of the Beagle*, 274-275) never mentions that FitzRoy paraded around carrying his Bible. A modern rendition states that FitzRoy "held up his Bible but implored the audience to believe God rather than man. He was shouted down" ["Wilberforce, Samuel" in Patrick H. Armstrong, *All Things Darwin: An Encyclopedia of Darwin's World*, 2 vols. (Westport, Conn.: Greenwood Press, 2007), 2:480f]. See also Nichols in *Evolution's Captain*, 318. These highly fictionalized accounts are added to further their strong Darwinian bias.

32 In a letter to Darwin, James Sulivan informed him of FitzRoy's declining health (*Correspondence*, 13:141-142). See also Mellersh, *FitzRoy of the Beagle*, 285-286, where he recorded the same diagnosis given by Sir Roderick Murchison (1792-1871), President of the Royal Geographical Society, and FitzRoy's wife. There was absolutely no mention about Darwin. Unfortunately, modern historians, wanting to elevate Darwinism, have created and propagated the scenario that Darwin's *Origins* was one of the causes that triggered FitzRoy's suicide. See Nichols, *Evolution's Captain*, 318f.

SACRED

TO THE MEMORY OF

ROBERT FITZROY

VICE ADMIRAL

BORN JULY 5, 1805

DIED AT NORWOOD, APRIL 30, 1865

"He that Believeth on the Son hath Everlasting Life."

John 3:36

"With great mercies will I gather thee. I hid my face from thee for a moment but with everlasting kindness will I have mercy on thee saith the Lord thy Redeemer."

Isa. 54:7,8

"All that are in the graves shall hear His voice and shall come forth; they that have done good unto the Resurrection of Life."

John 5: 28,29

## Chapter 6
# *Origin*—the 'sacred text' of Naturalism

John Kimball began his chapter on evolutionism in his popular biology textbook with this remark:

> In 1859, the British naturalist Charles Darwin published his *Origin of Species*. It has been said that his book ranks second to the *Holy Bible* in its impact on man's thinking.[1]

The comparison between these two books is highly appropriate for this chapter in that they both address two mutually exclusive religious viewpoints on origins: Naturalism and Supernaturalism. Even though Darwin wrote in the conclusion of his most enduring work: "I see no good reason why the views given in this volume should shock the religious feelings of anyone,"[2] readers, particularly

---

1 John Kimball, *Biology* (Reading, Mass.: Addison-Wesley, 1965), 539.
2 Charles Darwin, *On the Origin of Species* (1876), Vol. 16 of *The Works of Charles Darwin*, Paul H. Barrett and R.B. Freeman, eds. (London: William Pickering,

of today, should not be duped into seeing *Origin of Species* as solely a scientific manual. Rather, it should be viewed as a sacred writing which propagated a naturalistic theology. James Moore is absolutely correct when he stated:

> From start to finish *Origin* was a *pious work*: "one long argument" against creationist orthodoxy, yes, but equally a reformer's case for creation by natural law.[3]

## *VESTIGES*—A PRECURSOR OF *ON THE ORIGIN OF SPECIES*

*On the Origin of Species*—the encapsulation of Darwin's new gospel[4]—provided a naturalistic answer to the question: 'Where did we come from?' Fifteen years earlier, in 1844, Robert Chambers (1802-1871), a Scottish author and journalist, wrote "the first full-length presentation of an evolutionary theory of species in English."[5] He titled his book, *Vestiges of the Natural History of Creation*. Being the publisher of the famous *Chambers' Encyclopaedia*, he was accustomed to writing in a language that was easily understood by the common people.

Knowing that *Vestiges* would create a tidal wave of controversy and opposition, he decided to publish it anonymously. Periodicals of a religious or political nature rarely revealed the author's name but anonymity for historical or scientific writings was very unusual. Some sixty names[6] were put forth as possible authors, even Darwin's was suggested. Chambers' identity as the author of *Vestiges* was not revealed until 1885, fourteen years after his death. In the twelfth

---

1988), 439. Hereafter, it will be cited as *Origin* (1876).

3  James Moore, *The Darwin Legend* (Grand Rapids: Baker Books, 1994), 41. Italics not in the original.

4  "My species is all gospel." *Correspondence*, 4:140.

5  Tess Cosslett, ed., *Science and Religion in the 19th Century* (Cambridge: Cambridge University Press, 1984), 46.

6  James A. Secord, *Victorian Sensation: The Extraordinary Publication, Reception and Secret Authorship of the* Vestiges of the Natural History of Creation (Chicago: University Press of Chicago, 2000), 23. See also *Correspondence*, 3:181.

edition, a trusted friend, Alexander Ireland (1804-1894), officially ended the mystery.

From 1841 to 1844, Chambers secluded himself in the university town of St. Andrews, Scotland, where he read widely the sciences of the day, especially geology. Here, he devised an evolutionary theory of origins from the stars to mankind. "The probability may now be assumed that the human sprung from one stock which was first in a state of simplicity, if not barbarism."[7] Chambers was an advocate of Caucasian superiority. It was this race, he argued, that had successfully thrown off the shackles of barbarism; the other races, especially the blacks and Asians, were in a state of degeneracy.

As a deist,[8] Chambers portrayed God as a remote, almost impersonal, Great Watchmaker who set the natural laws in motion. Then, he ceased to be involved any longer in his creation. Natural laws, not God, were responsible for the great variety of living things. Charles Darwin, who already believed in evolutionism, recognized the book as "a grand piece of argument against immutability of species" and read it "with fear and trembling."[9]

References to a Divine Providence did not impress Adam Sedgwick, Professor of Geology at Cambridge and an ardent defender of Natural Theology. There is no doubt that this eminent scientist was Chambers' most outspoken critic. Sedgwick made the following scathing comments about *Vestiges* in a letter to Lyell on April 9, 1845:

> If the book be true, the labor of sober induction are in vain; religion is a lie; human law a mass of folly and a base injustice; morality is moonshine; our labors for the black

---

7  Robert Chambers, *Vestiges of the Natural History of Creation* (1844) [Repr. of the 1st ed., intro. by Gavin de Beer (New York: Humanities Press, 1969)], 305.

8  Secord, *Victorian Sensation*, 85, 171.

9  *Correspondence*, 3:258. In *More Letters*, 1:49, Darwin intimated in a letter to Joseph Hooker that the author was Robert Chambers.

of Africa were works of madmen; and men and women are only better beasts.[10]

Sedgwick's crowning insult was that the book of such despicable calibre had to be written by a woman![11]

In light of the above comment, it should be noted that Emma Martin (1812-1851), "the most celebrated propagandist for free thought and feminism,"[12] spoke out in defence of *Vestiges*. In 1844, this former Baptist turned atheist actually delivered a series of lectures at the Literary and Scientific Institution. During this Victorian era, many were aghast at Martin's confrontational and belligerent attitude. It was uttered by some that there was "something unnatural about an atheist in petticoats."[13] At the same time, free lectures against *Vestiges* were given in the newly-founded YMCA.

The cosmic scope of the book and its deistic view of God made it vulnerable for attack from various sectors of the academic world. Chambers' belief that life arose by means of spontaneous generation and then progressed from simple to more complex, of which mankind was the most complex, had little support in the scientific community of that day. The thought that mankind arose from a bestial past enraged many and they were quick to denounce the author as being "an atheist, shallow smatterer and credulous dupe."[14]

Chambers, wanting to garner the support of those committed to the Bible, felt that his evolutionary hypothesis could easily be harmonized with the scriptural account in Genesis. His pleas, as one would expect, fell on deaf ears. Nothing has changed. Some

---

10 John W. Clarke and Thomas M. Hughes, *The Life and Letters of the Reverend Adam Sedgwick*, 2 vols. (Cambridge: Cambridge University Press, 1890), 2:84.

11 A. Desmond and J. Moore, *Darwin: The Life of a Tormented Evolutionist* (New York: Warner, 1991), 322.

12 Secord, *Victorian Sensation*, 314.

13 Secord, *Victorian Sensation*, 317.

14 Milton Millhausen, *Just Before Darwin: Robert Chambers and* Vestiges (Middletown, Conn.: Wesleyan University Press, 1959), 4.

modern evolutionists are still endeavouring to weld naturalistic evolutionary assumptions within a biblical framework.

Two months after the publication of *Vestiges*, Joseph Dalton Hooker, a very close friend of Darwin, commented that "somehow the books [*sic*] looks more like a 9 days wonder than a lasting work."[15] Time certainly proved him wrong. Despite the caustic and condemnatory reviews, the book, being cheap and highly readable for the time, had brisk sales. There was a new monthly edition for the first four months. By 1854, there had been ten editions in the same number of years. In Britain alone, 24,000 copies had been sold.[16] This number did not include those sold in the U.S.A. where the book enjoyed a wide-spread popularity, nor those sold in Europe, where it was translated into Dutch and German.

Peter Brent's comment most aptly describes the reception of *Vestiges*: "Everywhere the book was being discussed, reviewed, supported, refuted *and above all read*."[17] The mystery of its authorship only intensified its mystique; Victorian notables like Queen Victoria, Benjamin Disraeli (1804-1881) and Florence Nightingale (1820-1910) availed themselves of this blockbuster. As expected, Charles Darwin read it. Though not publicly known as an evolutionist, he said of the book, "The writing and arrangement are certainly admirable but his geology strikes me as bad and his zoology far worse."[18]

In the 'Historical Sketch' of his *Origin*, Darwin pointed out that one of the major weaknesses of Chambers' theory was its lack of a mechanism of change. His references to mere "impulses"[19] as the driving impetus for change were totally inadequate. But his complete reliance on natural law, not divine intervention, in explaining

---

15 *Correspondence,* 3:103.

16 Janet Browne, *Charles Darwin: Voyaging* (New York: Knopf, 1995), 462.

17 *Charles Darwin: A Man of Enlarged Curiosity* (London: Heinemann, 1981), 398. Italics not in the original.

18 *Correspondence,* 3:108.

19 *Origin* (1876), xvii.

the development of life, gained Darwin's approval. Nevertheless, Darwin would have agreed with the assessment of a twenty-first-century Australian professor of zoology:

> Chambers' book was nineteenth-century pop-science: attractive, successful with a wide audience and quite inaccurate. Although condemned on all sides, the *Vestiges* nevertheless helped push 'the Species Question' to the center of the stage.[20]

The furore and hostility surrounding *Vestiges* was not lost upon Darwin. In that same year, he had expanded his theory on species into a larger essay from private correspondence written two years before. There is no doubt that Charles had every intention of publishing a book on this controversial subject; "yet he must have winced at the abuse raining down on the anonymous author. Was he in for the same treatment?"[21] Two questions plagued him: How could he avoid such ridicule and even resentment that was showered upon *Vestiges*? Secondly, no one could deny that *Vestiges* had captured the imagination of the British public, but how could he piggyback on this book's achievements in order to advance 'his own theory'?

Recognition as a competent biologist, especially within the scientific community, was vital. At that time, he had no formal training in biology.[22] To alleviate that serious deficiency, Charles began an arduous research project, lasting eight years (1846 to 1854), on a most unusual creature—the barnacle. His painstaking discoveries concerning this minuscule animal provided him with the

---

20 David Young, *The Discovery of Evolution*, 2nd ed. (Cambridge: Cambridge University Press, 2007), 101.

21 Janet Browne, *Darwin's Origin of Species: A Biography* (London: Atlantic Books, 2006), 52.

22 *Life and Letters*, 1:315. To Joseph Hooker, he admitted that he was a botanical ignoramous not knowing "a Daisy from a Dandelion" (*Life and Letters*, 1:312).

scientific credentials that he desperately needed. It has been estimated that he gathered 10,000 specimens! His perceptive insights
and deductions into this remarkable creature were recorded in
four volumes—two tomes on living ones and two shorter ones
on fossilized forms. "It established him as a zoological specialist,
and no longer just the geological expert. More important, it was
his licence to speak on species."[23] Furthermore, in honour of his
work on the barnacle and his geological studies in South America,
the Royal Society of London awarded Darwin the prestigious
Royal Medal in 1853.

These accolades could not have been attained without the assistance of a network of "collectors, friends, missionaries, business-
men, naturalists, mineralogists and shell collectors"[24] who sent
barnacle specimens to him from every corner of the world.
They could be sent in small glass jars or in pill boxes and arrived
at Darwin's estate in Downe, some 16 km (30 mi.) southeast of
London, via the ever-expanding railroad system and the new,
efficient postal service.[25]

Even though Darwin was sequestered in his home at Downe,
anyone throughout the world was only a letter away. This reliable
postal system became Darwin's lifeline for his barnacle research
and he fully utilized it. Through diplomacy and charm, he master-
fully persuaded others to send him barnacles and much-needed
information. To appreciate the scope of the letter writing of this
British scientist, one must understand that there were "14,000 letters
exchanged with some 1,800 correspondents over sixty-odd years."[26]

---

23 Desmond and Moore, *Darwin: The Life of a Tormented Evolutionist*, 409.

24 Rebecca Stott, *Darwin and the Barnacle* (London: Faber and Faber, 2003),
xxiv.

25 In 1837, Sir Roland Hill (1795-1879), a teacher and social reformer, pro-
posed a uniform penny post; he suggested "a prepaid flat rate for all letters under a
certain weight, no matter what the distance it was being sent" (Secord, *Victorian
Sensation*, 28). Initiated three years later, it became an instant success.

26 James R. Moore, "Darwin's Genesis and Revelations," *ISIS* 76 (1985), 573.

James Moore has charted the annual flow of letters to and from Darwin from 1821 until the time of his death. The year 1869 was the peak year with an incredible total of 800 letters—500 received and 300 sent![27] Just managing this voluminous mail was truly an astounding feat, not to mention that the data from these world-wide sources had to be collected and collated. Truly, the managerial skills of Charles Darwin were extraordinary.

Late in October 1854, Darwin "banished his barnacles either to the British Museum or back to their collectors across the world."[28] The fears and concerns, raised by Chambers' *Vestiges* eight years earlier, had been resolved to Darwin's satisfaction. Now he could focus his attention on the publication of his theory on species and ready himself for the inevitable spiritual warfare that was surely to ensue with the introduction of his evolutionary doctrine.

In the same year, England declared war on Russia in March and sent a quarter of a million men to the Crimean battlefront located on the Black Sea. Darwin's correspondence reveals that he, along with his countrymen, followed the war effort in the Crimea with much interest. Before the conflict ended two years later, such names as Balaclava and Inkerman (names of battles) were etched on the minds of the British public.

This war took the lives of nearly 20,000 soldiers—16,000 of whom died from diseases.[29] Florence Nightingale and a small corps of nurses whom she trained were dispatched in November 1854 to address this horrific problem. Following the war, she was instrumental in establishing nursing schools in Britain.

### DEATH CROSSES THE PATH OF A DEIST

In 1865, some six weeks or so after Joseph Hooker's father died, Charles wrote this letter of condolence to his close friend. His refer-

---

27 Moore, "Darwin's Genesis and Revelations," 574.

28 Stott, *Darwin and the Barnacle*, 240.

29 "Crimean War," on-line (en.wikipedia.org/wiki/Crimean_War), accessed March 6, 2008.

ence to the deaths of his own father who had died seventeen years earlier (1848), and his daughter, Annie, three years later (1851), showed that their memories were still uppermost in his thoughts.

> I do not think anyone could love a Father much more than I did mine & I do not believe three or four days even pass without my still thinking of him, but his death at 84 caused me nothing of that insufferable grief, which the loss of poor dear Annie caused. And this seems to me perfectly natural, for one knows for years previously that one's Father's death is drawing slowly nearer & nearer; whilst the death of one's child is a sudden and dreadful wrench.[30]

In the late spring of 1876 at the age of 67, Charles began to write his *Autobiography*.[31] His father was frequently mentioned; in one lengthy discourse, he recalled that his father had an extraordinary memory, "especially dates, so that he knew, when he was very old the day of his birth, marriage, and the death of a multitude of persons in Shropshire."[32] Even though Charles had been very dependent upon his father for medical and financial advice (Robert Darwin had been a shrewd investor), he never attended his father's funeral in 1848. As a thirty-nine-year-old adult, this was Darwin's first encounter with the death of a close family member. Darwin reasoned that he was much too ill to attend "his funeral or to act as one of his executors."[33] During the next year, he languished at Down House and had little or no contact with the outside world.

According to Darwin's reading list of books at that time, he mentioned three written by Francis William Newman (1805-1897).

---

30 *Correspondence*, 13:245.

31 "He completed the main text at Down about two months later. Over the next five years Darwin enlarged the manuscript by almost fifty per cent" [James Moore, "Of love and death: Why Darwin 'gave up Christianity'," in *History, Humanity and Evolution* (Cambridge: Cambridge University Press, 1989), 199].

32 *Autobiography*, 39. The discourse runs from 28-42.

33 *Autobiography*, 117. *Correspondence*, 4:183.

The one that gripped his attention was *Phases of Faith* (1850).[34] This was Newman's spiritual biography and, in  many ways, it paralleled Darwin's own spiritual quest.

At the age of sixteen, Newman was confirmed in the Anglican Church but he felt that the Bishop, "a *made-up* man and a mere pageant,"[35] had only one concern—that he be able to recite the prescribed creeds and catechism. There was no attempt to discover if young Newman actually *believed* these articles of faith.

The next year, Newman's first step toward fulfilling his dream of serving as a minister in the Anglican Church came when he was accepted as a theology student at Oxford University. Newman gladly subscibed to the *Thirty-Nine Articles*. After finding out that the majority of faculty members and students did not believe them, he could not contain his disappointment and dismay. Consequently, he felt "from his first day there, that the system of compulsory subscription was hollow, false and wholly evil."[36]

In 1830, four years after graduation, Newman, like Charles Darwin, went abroad. He went to India as a missionary assistant. After a three-year tenure, the mission sent him back to England to procure financial support. For some time, he had been wrestling with many areas within Christian theology. Was the Bible really the infallible and inerrant Word of God? He came to this conclusion: "The Bible was made for man, not man for the Bible."[37] Once having established this framework, he questioned the validity of the Trinity, more specifically the deity of Christ, and also the existence of eternal punishment. By examining them from a purely rationalistic point of view, he determined that these doctrines had to be rejected.

---

34 The other two were: *A History of the Hebrew Monarchy* (1847) and *The Soul, Its Sorrows and Aspirations* (1849).

35 Francis W. Newman, *Phases of Faith* (New York: Humanities Press, 1970), 2. Italics in the original.

36 Newman, *Phases of Faith*, 2.

37 Newman, *Phases of Faith*, 86.

The historicity of Genesis 3 also came into question. Like many young English scholars, he too had accepted the belief in uniformitarianism. "Geologists assured us, that death went on in the animal creation many ages before the existence of man… to refer the death of animals to the sin of Adam and Eve was evidently impossible."[38] This spiritual defection put an end to any service within the Anglican Church. He now turned to the Dissenting universities where he had a successful career teaching Latin. When Robert Chambers published his *Vestiges* in 1844, Newman resonated with this view on origins and became an ardent advocate of evolutionism.

In 1850, *Phases of Faith* came off the press; the story of Newman's spiritual journey struck a chord with many Britons. Over the next ten years, there were six editions. On a personal level, Newman's quest for meaning in life led him to become a committed Unitarianian, much like Emma Darwin. Charles Darwin was of a different opinion. Even though he identified very much with Newman's spiritual pilgrimage and had even recorded that *Phases of Faith* was an 'excellent' book, nevertheless, he fully agreed with his grandfather Erasmus Darwin that Unitarianism was nothing more than 'a feather bed to catch a falling Christian.' Deism was the only viable option and it must be embraced. Christianity had to be rejected once and for all.[39]

The death of his ten-year-old daughter, announced in the letter below to Emma, severely tested Darwin's deistic faith:

> I pray God Fanny's note may have prepared you. She [Annie] went to her final sleep most tranquilly, most sweetly at 12 o'clock today. Our poor dear dear child has had a very short life but I trust happy, & God only knows what miseries might have been in store for her.[40]

---

38 Newman, *Phases of Faith*, 68-69.
39 Desmond and Moore, *Darwin: The Life of a Tormented Evolutionist*, 378.
40 *Correspondence*, 5:24. It was dated April 23, 1851.

Known for her gregarious, buoyant disposition, Annie was his favourite child. During the arduous days of his barnacle research and writing, she would fuss over him continuously. She often accompanied him on his routine walks. Charles thoroughly enjoyed her companionship.

It was in July 1850 that Annie became ill. After months of various treatments and consultations, Charles decided to place her under the care of Dr. James Gully, a hydropathic physician whose treatment center was located at Malvern, some 240 km (150 miles) from the village of Downe. This was in March 1851. Charles himself had been one of his patients and thought that Annie might benefit from his water treatments. Three weeks later, Annie's health had deteriorated to such an extent that Charles was summoned to be with her. Leaving Emma at home, as she was in the final weeks of pregnancy, Charles rushed to be by her side. Through the last week of their daughter's life, Charles and Emma daily exchanged letters that reveal most poignantly their deep concern for their child but equally demonstrated the depth of their love for each other.

Shortly after Annie's death, Charles left Malvern for home. He entrusted the funeral arrangements to his sister-in-law, Fanny Wedgwood, the wife of Hensleigh, Emma's older brother. Distraught with grief, Charles never returned, not even to attend Annie's funeral service. As was noted earlier, such was the case with his father's death.

Seven days after his daughter's death, in the seclusion of Down House, Charles penned a memorial to his beloved child. He portrayed Annie "as a type-specimen of all the highest and best in human nature. Physically, intellectually and morally, she was perfect;... Annie did not deserve to die; she did not even deserve to be punished—in this world, let alone the next."[41]

He closed his tribute to his 'angelic' Annie with these words:

---

41 Moore, "Of love and death: Why Darwin 'gave up Christianity'," 218-220.

> We have lost the joy of the Household, and the solace of our old age:—she must have known how we loved her; oh that she could now know how deeply, how tenderly we so still & shall ever love her dear joyous face. Blessings on her.[42]

This memoir was one of despair. For Charles, as a Deist, there was no life beyond the grave; Annie was gone forever. "There was no straw to clutch, no promised resurrection. Christian faith was futile."[43] The celebration of Easter, a week before, with its message of hope beyond the grave was, from Darwin's perspective, meaningless and hollow. On her tombstone, there were no biblical passages, although that was almost a nineteenth-century tradition. Instead was the inscription: 'A dear and good child.'

Four months later, Henrietta (Annie's younger sister by two years) had a bedtime conversation with Emma about Annie's final resting place. On the basis of her Unitarian faith, Emma endeavoured to console her young daughter.

> Henrietta: "I am afraid of going to hell."
> I told her I thought Annie was safe in Heaven.[44]

Many years later, Henrietta made this most revealing comment:

> It may almost be said that my mother never really recovered from this grief. She very rarely spoke of Annie but when she did the sense of loss was always there unhealed.[45]

---

42 *Correspondence*, 5:542.

43 Desmond and Moore, *Darwin: The Life of a Tormented Evolutionist*, 384.

44 *Correspondence*, 5:543. Darwin's younger sister, Emily Catherine, wrote: "No little dear could have had a happier life except her health till her little innocent spirit was called above" (*Correspondence*, 5:30).

45 Henrietta Litchfield, ed., *Emma Darwin: A Century of Family Letters, 1792-1896*, 2 vols. (London: John Murray, 1915), 2:137.

**DOWN HOUSE** • This spacious house near the town of Downe was a sign of Darwin's aristocratic standing in British society.

## JOSEPH HOOKER—DARWIN'S FIRST CONVERT

On the summer eve of 1842, just after purchasing the Down estate with money borrowed from his father, Charles pencilled his first written account concerning his evolutionary theory, based on his notebooks of 1837 to 1839. Two years later, he produced (what is commonly called) 'The Sketch of 1844'; it is a more detailed account than the first and formed the basis for *Origin* some fifteen years later.

As a result of a reoccurring sickness[46] that started after he began work on 'his theory', Darwin felt his death was imminent. So, in a letter dated July 5, 1844, "as my most solemn and last request,"[47] he instructed his dutiful wife, Emma, how and from whom she should ask for assistance in having this work, which he felt would "be a considerable step in science,"[48] published. Lyell, Henslow and Hooker were mentioned as those who could be approached to help. Ten years later, on the cover of the same document, he wrote: "Hooker by far best man to edit Species volume."[49]

What happened in the ten-year interval that prompted Darwin to have such confidence in Joseph Hooker? During the 1830s, Lyell was undoubtedly the most influential person in Darwin's life, especially in the field of geology. Being twelve years his senior and regarded as one of Britain's leading scientists, Darwin saw Lyell more as a father figure and a wise counsellor rather than his equal. In 1844, Darwin never felt sufficiently at ease to share his radical thoughts on speciation with Lyell. But Joseph Dalton Hooker was different. Being eight years Darwin's junior and one who had always been exposed to the natural sciences through his father, Sir William Jackson Hooker (1785-1865)—a prominent botanist and first

---

46 See Ralph Colp, *To Be an Invalid* (Chicago: University of Chicago Press, 1977) for the best discussion on the causes of Darwin's illnesses. Dr. Colp, a physician, comes to the conclusion that Darwin's conversion to evolutionism and his pursuit of it are directly responsible for his illness.

47 *Correspondence*, 3:43.

48 *Correspondence*, 3:43.

49 *Correspondence*, 3:44, footnote 10.

Director of the Royal Botanical Gardens at Kew—he had very much wanted to emulate Charles Darwin. In 1839, this opportunity presented itself: as a recent graduate in medicine from the University of Glasgow, he was asked to be the naturalist on HMS *Erebus* to Antarctica. On this trip, from 1839 to 1843, "Darwin's *Journal* was his constant companion."[50]

On January 11, 1844, Darwin dropped a bombshell onto the lap of Hooker—an acquaintance that he had only known for a month—when he wrote:

> I determined to collect blindly every sort of fact, which could bear any way on what are species... At last gleams of light have come and I am almost convinced (quite contrary to the opinion I started with) that species are not (*it is like confessing a murder*) immutable.[51]

Darwin, revealing not only his anxiety but his innermost struggle concerning the species question, made Hooker his first confidant. The startling words 'like confessing a murder' have caused much bewilderment to Darwinian scholars. The fact that the words are stated parenthetically—a typical method used by Darwin to show his strong feelings of repression or suppression[52]—only adds to the problem. Frank Burch Brown has offered the best analysis in light of Darwin's evolutionary commitment of the late 1830s by seeing that "the implied victim to the 'murder' was the God of orthodox theism."[53]

Guilty of deicide! How could Charles ever face his peers, let

---

50 Geoffrey West, *Charles Darwin: A Portrait* (New Haven: Yale University Press, 1938), 194. From 1848 to 1851, Hooker also went to the Himalayas and India.

51 *Correspondence*, 3:2. Italics not in the original.

52 Colp, *To Be an Invalid*, 29. For a fuller treatment by the same author, see "'Confessing a Murder': Darwin's first Revelation about Transmutation," *ISIS* 77 (1986), 9-32.

53 *The Evolution of Darwin's Religious Views* (Macon, Georgia: Mercer University Press, 1986), 19.

**JOSEPH DALTON HOOKER (1817-1911)** • Hooker promoted evolutionism
within the scientific communities in Britain.

alone the people of Downe? Here he had become "an establish-
ment figure in his own right, a pillar of the parish."[54] Even though
all of his children were christened in the Anglican church in Downe
and he gave generously, particularly to special needs, he had come

54 Moore, *The Darwin Legend*, 39.

to visualize himself as a spiritual quisling—"the Devil's Chaplain."[55] In the 1840s, evolutionism was viewed by Victorian England as "false, foul, French, atheistic, materialistic and immoral."[56] Darwin took this vitriolic characterization against his belief system to heart. It instilled within him such fear and trepidation that he purposely withheld publishing his book on origins for the next fifteen years.

Besides observing the religious overtones of the confession, one must not fail to see that Darwin now recognized in Hooker, as a rising botanist, an equal in whom he could not only confide, but also one whom he could use to assist him in providing information to develop an intellectual edifice upon his accepted assumptions. Hooker was more than willing to be "Darwin's personal sounding board"[57] in giving shape to his species work from 1844 to 1859.

Darwin, as we have noted, was a prolific letter writer. This is most evident when one examines his correspondence with Joseph Hooker. Hooker, his closest friend outside his immediate family, received some 1,394 letters—ten per cent of the letters that Darwin penned over a forty-year period.[58] The Master of Down was the first to recognize his dependence on Hooker's expertise in the natural sciences. Hooker's willingness "to be pumped"[59] by Darwin, who had endless questions, was indispensable in developing the background material for *Origin*. Darwin's feeling of gratitude was best expressed in a letter to his most trusted friend when he wrote:

> You are the one living soul from whom I have constantly received sympathy. Believe me that I never forget for even a minute how much assistance I have received from you.[60]

---

55 *Correspondence*, 6:178.

56 Desmond and Moore, *Darwin: The Life of a Tormented Evolutionist*, xviii.

57 Barry G. Gale, *Evolution Without Evidence: Charles Darwin and the Origin of Species* (Albuquerque: University of New Mexico Press, 1982), 55.

58 Moore, "Darwin's Genesis and Revelations," 576. The next closest was Thomas Huxley with 285 letters.

59 *Life and Letters*, 1:387.

60 *Life and Letters*, 1:494f.

This intimate and amiable relationship amply explains why Darwin, in 1855, instructed Emma that Hooker was indeed the best person to consult if she needed help in publishing his 'Sketch of 1844.' But despite his indebtedness to Hooker, Darwin was deeply disturbed by the fact that this eminent scientist and close friend was not yet convinced that 'his theory' was valid.

Finally, in a letter dated August 11, 1858 to Asa Gray (1810-1888), Darwin announced that Hooker had been converted.[61] This confession to Asa Gray, a correspondent with Darwin for almost two decades was highly significant. Gray was "recognized as America's foremost botanist and in 1842 was appointed professor of natural history at Harvard."[62] In writing to Alfred Wallace (1823-1913) some eight months later on April 6, 1859, Darwin could hardly contain himself, and the glee now seems to jump off the page as he declared:

> I forgot whether I told you that Hooker, who is our best British botanist and perhaps the best in the world, is a *full convert* and is now going immediately to publish his confession of faith; and I expect daily to see proof-sheets.[63]

Also, almost two months to the day before *Origin* came off the press on November 1859, Darwin wrote to his cousin William Darwin Fox and did not fail to mention the recent conversion of Hooker, then Assistant Director of the Royal Botanical Gardens at Kew. Despite this news, Darwin had his doubts whether the complimentary copy of *Origin* soon to be sent would convince his sceptical cousin to accept his evolutionary theory. Darwin wrote of Lyell: "He has read about half of the volume in clean sheets and gives me great *kudos*. He is wavering so much about the immutability of

---

61 *Life and Letters*, 1:491.

62 William E. Phipps, *Darwin's Religious Odyssey* (Harrisburg: Trinity Press International, 2002), 59.

63 *More Letters*, 1:119. Italics not in the original.

**ASA GRAY (1810-1888)** • As a theistic evolutionist, Gray effectively promoted Darwin's *Origin* in America.

species that I expect he will come around."[64] Interestingly, it never happened. In 1863, when Lyell wrote *The Antiquity of Man*, he continued to believe that a great chasm existed between animals and humans. His failure to adopt an evolutionary position was a great

---

64 *Life and Letters*, 1:522. Italics in the original.

disappointment to Darwin. "Despite all their friendly connections, Darwin never really forgave him for betraying his hopes."[65]

Why were Darwin's closest associates having difficulty in accepting his theory? Possibly, the answer lies in a statement to Hooker a month or so after the publication of *Origin*:

> It is an old and firm conviction of mine that the Naturalists who accumulate facts and make partial generalizations are the real benefactors of science. Those who merely accumulate facts I cannot very much respect.[66]

In light of the status and overall acceptance of evolutionism in our present society, Darwin's admission that his theory was based on 'partial generalizations' and circumstantial evidence is highly significant. The proponents of evolutionism who are vociferously claiming that it is an undeniable fact should take note. Furthermore, their dogmatism is again challenged when one considers that Darwin's 'generalizations' may well be based on a number of spurious assumptions. High school textbooks have only heightened the problem by including chapter headings on the 'proofs' of evolutionism.

## SPREADING 'THE GOSPEL' OF *ORIGIN*

On November 24, 1859, John Murray Publishing Company printed *On the Origin of Species by means of Natural Selection or the Preservation of Favored Races in the Struggle for Life.* Darwin referred to it as the most important work of his life. In order to divert his attention from his species theory, Darwin installed a pool table in a room next to his study. In a letter to John Murray (1808–1892), when speaking of his book, he wrote: "I am *infinitely* pleased & proud at the appear-

---

65 See Janet Browne, *Charles Darwin: The Power of Place* (New York: Knopf, 2002), 2:220.

66 *Autobiography*, 122.

ance of my child."[67] After the publication of *Origin*, "Darwin's reputation as a biologist superceded his renown as a geologist."[68]

Nine months before Murray agreed to publish *On the Origin of Species*, he had been informed that the book made absolutely no references to the evolutionary development concerning mankind or to the book of Genesis.[69] But, as any publisher would do, he approached two associates, George Frederick Pollock (1786-1872) and Whitwell Elwin (1816-1900), to review Darwin's manuscript. On May 3, 1859, Rev. Elwin sent his assessment of *Origin* in a letter to Murray, which was later forwarded to Darwin.

Whitwell Elwin, the Rector of Booton and also editor of the *Quarterly Review*, believed that the book was much too speculative and should not be published in its present form:

> It seemed to me that to put forth the theory without evidence would do grievous injustice to his views, & to his twenty years of observation & experiment. At every page I was tantalized by the absence of the proofs... It is to ask the jury for a verdict without putting the witnesses into the box.[70]

Elwin had nothing but the highest regard for Charles' former publications but he recommended that Darwin, as one of England's foremost scientists, should publish his research on pigeons and abandon his evolutionary material for the moment. Elwin thought his research on pigeons was "curious, ingenious, & valuable in the highest degree... Everybody was interested in pigeons."[71] Sir J. William Dawson (1820-1899), who was principal of McGill

---

67 *Correspondence*, 7:365. Italics in the original.

68 Sandra Herbert, *Charles Darwin, Geologist* (Ithaca, New York: Cornell University Press, 2005), xvi.

69 *Correspondence*, 7:270. See Browne, *Charles Darwin: The Power of Place*, 2:75.

70 *Correspondence*, 7:289.

71 *Correspondence*, 7:289-290.

University in Montreal, Canada, for thirty-eight years and North America's leading creationist, voiced the same concern. In the *Canadian Naturalist and Geologist* (1860), he wrote in a review that all Darwin proved in his book was that "a pigeon with all its varieties, was still a pigeon, and, according to our author's own conclusive argumentation, a rock-pigeon."[72]

George Pollock took a different stance. He argued that not only should the book be published but that Murray should print 1,200 copies instead of the original 500. Following Pollock's advice, Murray printed 1,250 green-covered volumes of *On the Origin of Species*.

This first edition was sold in one day. Ian Taylor convincingly argues that it was bought up at the dealer's auction by an agent of Lyell or Hooker a week or so before the official date of publication[73] and then sent out as complimentary copies to individuals in influential positions. Within a ten-year period and by the fifth edition, only 8,000 copies had been sold in Britain which had a population of some thirty million people.[74] But at the same time, "Darwin's name and the terms of Evolution and Natural Selection had been established firmly in the minds of the public at large."[75] If so few books were sold, how then did Darwin's name and these terms become household words?

Apart from the more literate, few people of the last century, even as today, waded through the 500 pages of *Origin*. Professor Ellegård

---

72 J.W. Dawson, "Review of 'Darwin on the Origin of Species by means of Natural Selection'" in *The Canadian Naturalist and Geologist* 5 (1860), 109. See author's *The Faces of Origins*, 92,96-98.

73 Ian Taylor, *In the Minds of Men: Darwin and the New World Order* (Toronto:TFE Publishing, 1984), 359-360. See also Morse Peckham, ed., *Charles Darwin, The Origin of Species: A Variorum Text* (Philadelphia: University of Pennsylvannia Press, 1959), 16, 20.

74 Alvar Ellegård, *Darwin and the General Reader: The Reception of Darwin's Theory of Evolution in the British Periodical Press 1859-1872* (Goteborg: Acta Universitis Gothenburgensis, 1959), 383. The exact number published through the first five editions was 9,750. See a chart in Peckham, *Charles Darwin, The Origin of Species: A Variorum Text*, 24.

75 Ellegård, *Darwin and the General Reader*, 43.

who has carefully examined 115 British periodicals of the 1860s, including 45 of a religious nature, makes a strong case that it was these accessible reading materials which publicized Darwin's ideas. But the areas most responsible for paving the way for the acceptance of evolutionism by the mid-nineteenth century were two: the influence of German biblical criticism and the scientific societies.

In 1846, the first real impact of biblical criticism—a movement started by German intellectuals to remove the supernatural element from the Bible—was felt in England when the *Life of Jesus* (1835) by David Strauss (1808-1874) was translated into English. This German author stripped away any vestige of supernaturalism from the Scriptures; the miraculous, especially the deity of Jesus Christ, was questioned. Startled by Strauss' moralistic Jesus but realizing the radical tendencies of some of the continental scholars, the English people who gave mere lip service to the Bible were shocked when seven liberal British authors—six Anglican clerics and one amateur geologist—published *Essays and Reviews* (1860). They were denounced as "Seven against Christ."[76] Their work declared the Bible to be "a collection of documents written by fallible human historians not as the directly inspired, literally true, Word of God."[77]

*Essays and Reviews*—a public denial of the historical value of the Bible—was in reality a true barometer of the religious sentiment of the Anglican hierarchy and the leading British intellectuals. The ultimate triumph for the liberal segment of British society came in 1896 when Frederick Temple (1821-1902), one of the contributors of *Essays and Reviews*, was appointed Archbishop of Canterbury—the highest position in the Anglican Church.

J.W. Bengough (1851-1923), the editor of a Canadian weekly newspaper, *Grip*, penned a satirical poem called "The Higher Criticism." His personal insights are quite revealing:

---

76 Desmond and Moore, *Darwin: The Life of a Tormented Evolutionist*, 500.
77 Cosslett, *Science and Religion in the 19th Century*, 12.

**DAVID STRAUSS (1808-1874)** • *The Life of Jesus, Critically Examined*, written by Strauss, a German 'higher' critic, denied the miraculous elements in the gospels. He saw them as being mythical in character. Strauss' book was published in English thirteen years before Darwin's *Origin* and exerted considerable influence, even among the Anglican clergy. German Higher Criticism was one of the key elements that provided fertile ground for the acceptance of the ideas that Darwin was to bring forward with the publication of his *Origin*.

**FREDERICK TEMPLE (1821-1902)** • Temple became Archbishop of Canterbury in 1896. This appointment signalled the triumph of Naturalism in the Anglican Church in England.

I saw a Higher Critic looking scholarly and cool.
As he stood beside the portals of the new Negation School;
And as I passed he stopped me by a motion of his hand.
Saying, "Pray don't look so much at ease—you do not
　understand."...

"In short, my simple brother, you must not feel so sure;
The Book you think inspired is only Jewish literature.
Its authorship, chronologies and dates are quite astray;
You must wait and hear what future excavations have to
　say..."

My dear old mother, dead and gone, was a Higher Critic
　too;
This Book was hers—she loved it, and she knew it
　through and through.
She told me 'twas from God direct, and she'd no doubt at
　all
The Patriarchs had really lived, as well as John and Paul.

She told me how the world was made, and all about the
　Flood,
And the Israelites were saved by the sprinkling of the blood.
She wasn't very learned, she did not know much Greek.
And of 'tentative suggestions' I never heard her speak.

But she was a Higher Critic of the very highest kind—
She searched the Scriptures daily the pearl of price to find;
She caught their inner spirit—which some Higher
　Critics miss—
And Christ was formed within her and filled her soul
　with bliss.[78]

---

78 J.W. Bengough, *Motley: Verses Grave and Gay* (Toronto: William Briggs, 1895),
163-166.

**'THE EVOLUTION OF MAN—AND WOMAN'** • Editor J.W. Bengough took delight in poking fun at evolution (*Grip*, March 1887).

*Essays and Reviews* caused a greater furore than Darwin's *Origin*. 11,000 Anglicans signed a declaration denouncing this German heresy. Much to the amazement of the essayists, 22,000 copies sold in two years.[79] As a means of comparison, it took two decades for Darwin's *Origin* to equal that number. Not only did *Essays and Reviews* shatter what little confidence the general public had in the biblical record, but it made *Origin* more appealing.

All that was needed was to give the public more exposure to Darwin's generalizations under the guise of science. It was the British scientific societies that would fulfil this role. Darwin's 'missionaries'—men like Thomas Huxley, Herbert Spencer (1820-1903)[80] and, of course, Joseph Hooker—were able to control the

79 Desmond and Moore, *Darwin: The Life of a Tormented Evolutionist*, 500.
80 Biologist Herbert Spencer coined the term "survival of the fittest" (*Principles*

executive offices of the scientific organizations such as the Royal Society of London and the British Association for the Advancement of Science (BAAS). In the latter, both Hooker and Huxley, in 1866 and 1870 respectively, served as president.

The mandate of the BAAS was to bring science to the general public. By avoiding London and going to various locations throughout the country, the leaders saw the attendance at their meetings go from 1,300 in 1830 to 2,500 in 1870.[81] They deliberately arranged to have the presidential inaugural address to be given in August when there was a lull in the news so that the newspapers would give maximum coverage to what became known as "the scientific event of the year."[82] From 1863 on, the presidential address was printed in its entirety. Hooker, in his 1866 address, was the first to centre his speech on Darwin's theory. Is it any wonder that by the 1870s English society was familiar with, and generally receptive to, evolutionism?

Certainly, with such devoted followers and press exposure, one of the goals of the father of evolutionism, as stated in his *Origin*, had been easily attained.

> I by no means expect to convince experienced naturalists whose minds are stocked with a multitude of facts all viewed, during a long course of years, from a point of view directly opposite from mine. ...but I look with confidence to the future,—*to young and rising naturalists.*[83]

By 1870, Charles Darwin was amazed at how quickly the British people accepted his theory and he had reason to be confident that

---

*of Biology* (1864), Vol. 1, 444: "This survival of the fittest, which I have here sought to express in mechanical terms, is that which Mr. Darwin has called 'natural selection,' or the preservation of favoured races in the struggle for life").

81 Ellegård, *Darwin and the General Reader*, 64.

82 Ellegård, *Darwin and the General Reader*, 65.

83 *Origin* (1876), 440. Author's italics.

evolutionism would captivate the keen minds of young scientists. But why? The biblical supernatural worldview upon which British science was built had been totally eroded away. The 'gospel' of Naturalism, with its Enlightenment roots in the previous century, had supplanted it and, by Darwin's time, had provided the ideal climate for *Origin* to emerge.

*On the Origin of Species* with its Natural Selection—the piloting force of life—was the death-knell to Natural Theology. The three quotations at the beginning of Darwin's book (only two in the first edition) dedicated to William Whewell (1794-1866), Joseph Butler (1692-1752) and Francis Bacon (1561-1626), all recognized as advocates of Natural Theology, did little to delude the discerning reader of the 1860s—or those of today. Darwin was totally opposed to Natural Theology; he felt that Natural Selection (often capitalized to show its supernatural qualities) had the capability of engineering all stages of the development of life. The intervention of any divine Person or Force was totally unnecessary.

Was Darwin replacing God with another god, namely Natural Selection? This issue of the deification of Natural Selection caused a heated disagreement between Darwin and Lyell. Darwin stated: "Yet in the same manner, as the architect is the *all*-important person in a Building, so is Selection with organic bodies."[84] Lyell replied that all the creative and intelligent attributes that Darwin had ascribed to Natural Selection would logically lead to its deification.[85] Two days later in an uncharacteristic curtness, Darwin inferred that Lyell's adherence to William Paley's *Natural Theology* was the sole reason for his unwillingness to see that he had not deified Natural Selection. Darwin ended the discussion with this abrupt comment: "But we shall never agree, so do not trouble yourself to answer."[86]

Let one excerpt from *Origin* illustrate the creative power of Natural Selection and also note its divine attributes, which are

---

84 *Correspondence*, 8:254. Italics in the original.
85 *Correspondence*, 8:255.
86 *Correspondence*, 8:258.

placed inside the brackets by the author:

> It may metaphorically be said that Natural Selection is
> daily and hourly scrutinizing, throughout the world, the
> slightest variations [omnipresence]; rejecting those that are
> bad, preserving and adding up all that are good [omni-
> science]; silently and insensibly working, *whenever and*
> *wherever opportunity offers*, at the improvement of each
> organic being in relation to its organic and inorganic con-
> ditions of life [omnipotence]. We see nothing of these slow
> changes in progress, until the hand of time has marked the
> lapse of ages and then so imperfect is our view into long-
> past geological ages that we see only that the forms of life
> are now different from what they formerly were.[87]

Its ability to scrutinize and select what was most beneficial was a
feature reminiscent of William Paley's beneficent God of *Natural*
*Theology*. The divine-like capabilities of Natural Selection had
been a concern for Darwin. Janet Browne made these insightful
observations:

> Natural Selection is not self-evident in nature nor is it the
> kind of theory one can say 'look here and see.' Darwin had
> no crucial experiment that conclusively demonstrated
> evolution in action. …Like Lyell in his *Principles of Geology*,
> he had to rely on an analogy between what was known and
> what was not known. He depended upon probabilities.[88]

In mid nineteenth-century England, and even more prevalent
today, was the lack of understanding that a new theological framework, and its accompanying underlying assumptions, had been
established. In reality, Darwin's 'scientific' research occurred within

---

87 *Correspondence*, 16:68f. Italics in the original.
88 Browne, *Darwin's Origin of Species: A Biography*, 69.

a new framework of which he was unaware. First, there was the doctrine of uniformitarianism which provided the necessary *time* for evolutionism to occur, and second, the principle of gradualism. Everything, he postulated, developed little by little over a long period of time. "Time, chance and reproduction ruled the earth. Struggle too."[89] Neither of these two can be demonstrated scientifically but rather are faith positions.

To complicate matters for Charles, in the second edition which was printed two months after the first, he added the little phrase 'by the Creator' in the concluding paragraph. The new version read:

> There is grandeur in this view of life, with its several powers having been originally breathed *by the Creator* into a few forms or into one…from so simple a beginning endless forms most beautiful and most wonderful have been and are being evolved.[90]

Privately to Hooker three years later, Darwin expressed his regret for this in very strong language:

> I have long regretted that I truckled to public opinion and used the Pentateuchal term of creation, by which I really meant 'appeared' by some wholly unknown process. It is mere rubbish, thinking at present of the origin of life, one might as well think of the origin of matter.[91]

The question that is difficult to answer is: Was this 'truckling' a lack of courage on Darwin's part to go all the way and declare himself to be an avowed atheist who was completely committed to a naturalistic worldview, or was there a deep-seated desire to recognize a Supreme Being?

---

89 Browne, *Darwin's Origin of Species: A Biography*, 67.
90 *Origin* (1876), 446f. Italics not in the original.
91 *Correspondence*, 11:278; *Life and Letters*, 2:202f.

# Chapter 7
# On God and humankind

The beauty of his character, charm of his manners, and heartiness of his friendship, were such as is most rarely met with, and, as a relation of mine lived near him for a short time observed, no one could know, and not love him.[1]

Rev. John Brodie Innes (1817-1894), Darwin's minister and intimate friend of the family for a period of forty-five years (1848 to 1893) penned these words in a private letter. The rural folk of Downe could well have resonated with Innes' assessment of Charles Darwin.

In 1848, during Rev. Innes' first year at St. Mary's, he felt there was a desperate need to provide social assistance to the people of Downe. He approached Charles and asked him to serve "as treasurer of the Coal and Clothing Club that provided welfare for the

---

1 Robert M. Stecher, "The Darwin-Innes Letters: The Correspondence of an Evolutionist with His Vicar," *Annals of Science* 17 (1961), 255.

needy in their village."[2] Monies were collected monthly and distributed to meet winter necessities. Emma gave out food vouchers that could be redeemed at the local bakery.

Two years later, the Downe Friendly Society was established. Once again, Charles agreed to be the treasurer. This program "provided insurance to cover against the loss of work, illness or funeral charges."[3] Darwin's humanitarianism also extended to the treatment of animals. By joining the Society for the Prevention of Cruelty to Animals, he added his opposition to cock-fighting and badger-baiting. He even took a sheep farmer to court for allowing his animals to starve to death.

## IS THERE A GOD?

Unlike Emma and the children who went to church regularly, Charles rarely attended. But he served on the parish council and was quite liberal in his contributions to the needs of the church in Downe. Even more remarkable was his intimate relationship with John Brodie Innes. In regards to theological issues, and especially origins, they had this gentlemen's agreement. "You are a theologian, I am a naturalist, the lines are separate. I endeavor to discover facts without considering what is said in the Book of Genesis. I do not attack Moses and I think Moses can take care of himself,"[4] Darwin quipped.

Even though this compartimentalization of science and religion may have been successful in maintaining a harmonious friendship, it was based on a common fallacy. In reality, when Darwin was wrestling with the problem of the origin of species, he was engaged in a religious endeavour, not a scientific one. Had Darwin and Innes recognized this crucial fact, would it have marred their relationship?

---

2   William E. Phipps, *Darwin's Religious Odyssey* (Harrisburg: Trinity Press International, 2002), 46. See *Correspondence*, 8:231.

3   Janet Browne, *Charles Darwin: The Power of Place* (New York: Knopf, 2002), 2:453.

4   *Life and Letters*, 2:82.

**ST. MARY'S ANGLICAN CHURCH, DOWNE** • This small village church in the County of Kent, outside London, was where the Darwin family attended.

In the early years of the 1860s after the publication of *Origin*, Darwin discussed with Asa Gray, in their numerous letters, his theological concerns about the design argument. This prominent American botanist was able to garner support for Darwin's evolutionary doctrine, especially in the American scientific community. As editor of the *American Journal of Science and Arts*, he wrote a highly positive review of *On the Origin of Species*.

In his three-part series on *Origin* in the July 1860 issue of *Atlantic Monthly*, Gray was able to outline his own theistic evolutionary position.[5] Doubtless, Darwin was elated; two months later, he complimented Gray by stating: "The two last essays are by far the best Theistic essays I ever read."[6]

It was within this bond of mutual respect and admiration that Darwin felt that he could bare his soul concerning the issue of design within the natural world. One candid reply was very revealing:

---

5   Theistic evolutionism maintains that God orchestrated the evolutionary development over eons of time. See Phipps, *Darwin's Religious Odyssey*, 59.

6   *Correspondence*, 8:388.

My dear Gray,

I thank you for two letters. ...Yesterday I read with care the third Article [that Gray wrote]; & it seems to me, as before *admirable*. But I grieve to say that I cannot honestly go as far as you do about Design. I am conscious that I am in an utterly hopeless muddle. I cannot think that the world, as I see it, is the result of chance; & yet I cannot look at each thing as a result of Design.[7]

Note that Darwin would not entertain the thought of atheism. He admitted to Gray that the complexity and beauty of life demanded "a First Cause."[8] Three years before his death, Darwin stated very emphatically, "In my most extreme fluctuations I have never been an Atheist in the sense of denying the existence of a God."[9]

But even though the squire from Down House admitted there was some Supreme Being, he still continued to be in a deep quandary. What is the relationship between this Master-Designer and the obvious appearance of design and order within the natural world? Darwin wrote:

If I was to say that I believed this [humankind designed], I should believe it in the same incredible manner as the orthodox believe the Trinity in Unity.—You say that you are in a haze; I am in thick mud;—the orthodox would say in fetid abominable mud. ...yet I cannot keep out of the question.[10]

Darwin's resolution of 'this muddle' was: A First Cause or, better still, a Deistic Being, set in motion the operative laws within the

---

7 *Correspondence*, 8:496. Italics in the original. See also *Correspondence*, 8:275. The letter ended: "Your muddled & affectionate friend."

8 *Correspondence*, 8:389.

9 *Life and Letters*, 1:275.

10 *Correspondence*, 9:369.

universe and then withdrew, having no further involvement. From that moment on, these established laws brought forth the order and design within the realm of nature through a random evolutionary process over eons of time.

## DOES MANKIND HAVE A BESTIAL PAST?

"Much light will be thrown on the origin of man and his history,"[11] was the only reference in *Origin* that Darwin made to mankind's place in the evolutionary scheme of life. In a letter to Alfred Wallace in 1857 (two years prior to the publication of *Origin*) he wrote: "You ask whether I shall discuss 'man.' I think I shall avoid the whole subject, as it is so surrounded with prejudices; though I fully admit that it is the highest and most interesting problem for the naturalist."[12]

His notebooks, written some twenty years earlier, reveal the same perspective. After his acceptance in 1837 that all species were mutable (and, accordingly, his conversion to evolutionism), he "could not avoid the belief that man must come under the same law."[13] In 1871, Darwin finally addressed this question in his book, *The Descent of Man, and Selection in Relation to Sex.*

Its theme was "the unity of man with the rest of the evolving world of animate life on the earth."[14] Twelve years had elapsed since the publication of *Origin* but the question persisted in the minds of Victorians: Were their ancestors Adam and Eve or were they a product of Natural Selection? Darwin's *Descent of Man* stated categorically that humankind had a bestial past.

In 1832 on the *Beagle* trip, at the tip of South America, Charles had encountered the native people of Tierra del Fuego for the first time. This frightful experience—proof positive to him that man

---

11 *Origin* (1876), 445.

12 *Life and Letters*, 1:467.

13 *Autobiography*, 130.

14 John Durant, "The Ascent of Nature in Darwin's *Descent of Man*" in David Kohn, ed., *The Darwinian Heritage* (Princeton: Princeton University Press, 1985), 294.

evolved—was never forgotten. In the second to last paragraph of *Descent of Man*, Darwin painted, in graphic terms, a vivid description of these Fuegians:

> But there can hardly be a doubt that we are descended from barbarians. The astonishment which I felt on seeing a party of Fuegians on a wild and broken shore will never be forgotten by me, for the reflection at once rushed into my mind—*such were our ancestors.* ...He who has seen a savage in his native land will not feel much shame, if forced to acknowledge that the blood of some more humble creature flows in his vein.[15]

These Fuegians—hideous, half-naked savages—evoked in Darwin a sense of consternation and disgust. But as *our ancestors*, they were at some point between an Englishman and an ape.

In the introduction to the *Descent of Man*, Darwin used, for the first time, the term, "evolution."[16] Since mankind and animals had a common ancestry, they must bear a similarity of nature. The difference in intelligence "between man and the higher animals, great as it was, certainly was one of degree and not of kind."[17] Only through the power of Natural Selection, did humans have the capability of abandoning their barbaric state and developing into civilizations, such as ancient China and Egypt. Darwin cited the United States as the best example of a modern nation and its people that had been guided by Natural Selection into nationhood. "For the more energetic, restless and courageous men from

---

15 Charles Darwin, *The Descent of Man, and Selection in Relation to Sex*, Vol.21 of *The Works of Charles Darwin*, Paul H. Barrett and R.B. Freeman, eds. (London: Pickering, 1988), 644. Italics not in the original. Hereafter, it will be cited as *Descent of Man*.

16 *Descent of Man*, 4.

17 *Descent of Man*, 130.

all parts of Europe emigrated during the last ten or twelve generations to that great country, and have succeeded best."[18]

*Descent of Man* was an instant success. Two months after its initial
printing in February 1871, 4,500 copies had been sold.[19] Despite
the ongoing turmoil in Europe caused by the Franco-Prussian War,
Darwin's second classic was translated into Dutch, French, German
and Russian. Much to the delight of Darwin, he received a substantial compensation for his efforts from John Murray, his publisher.

This book was Darwin's first public pronouncement on his personal views concerning religion, and more specifically—God. He
was indebted to Edward Burnett Tylor (1832-1917) for his book,
*Primitive Culture* (1871). In a letter to this father of modern anthro-
pology, Darwin was most complimentary: "It is wonderful how
you trace animism from the lower races up to the religious beliefs
of the higher races."[20] Following Tylor's conjectures, Charles believed
that all religions developed along this evolutionary pattern: "fetish-
ism, [then] polytheism and ultimately into monotheism."[21]

The belief in spirits, particularly among the uncivilized, has always
been recognized as universal in its scope but "the idea of a universal
and beneficent Creator does not seem to arise in the mind of man,
until he has been elevated by long-contained culture."[22] Darwin's
theological position here was totally opposed to the biblical position
that mankind was created with the innate understanding of God.[23]
He contended that *humans* devised the concept of "an all-seeing
Deity,"[24] which had a profound effect on their moral development
as they became more sophisticated both intellectually and culturally. Societies marked by such virtues as loyalty, courage, fidelity

---

18 *Descent of Man*, 147.

19 A. Desmond and J. Moore, *Darwin: The Life of a Tormented Evolutionist* (New
York: Warner, 1991), 579.

20 *Life and Letters*, 2:331.

21 *Descent of Man*, 100.

22 *Descent of Man*, 637.

23 Romans 1:18-32.

24 *Descent of Man*, 637.

and care and concern for others "would be victorious over most tribes; and this would be Natural Selection."[25]

Darwin made the following revealing statement in a letter to William Graham (1839-1911), the author of *The Creed of Science: Religious, Moral, and Social* (1881) in which he, in his own mind, had reconciled the belief in God *and* evolutionism:

> Nevertheless, you have expressed my inward conviction, though far more vividly and clearly than I could have done, that the Universe is not a result of chance. But then with me the horrid doubt always arises whether the convictions of man's mind, which has been developed from the mind of lower animals, are of any value or at all trustworthy. Would anyone trust the convictions of a *monkey's mind* if there are any convictions in such a mind?[26]

Therein lay Darwin's dilemma! By presupposing, as he did, that man's religious ideas evolved and, thus, were naturalistic in origin, how could one ever deem them to be trustworthy? What profound truths of a spiritual nature could the mind of a being—half-man, half-monkey—have? Since, to his way of thinking, religions and the idea of God were the culmination of an evolutionary process, Darwin had every reason to be sceptical.

Surprisingly enough, while penning *Descent of Man*, Charles was corresponding with Admiral James Sulivan who had been a Lieutenant on the *Beagle*. As an evangelical, Sulivan believed that the transforming gospel of Jesus Christ could make an impact on the lives of the Fuegians. Darwin disagreed. 'These savages,' he argued from the first time that he saw them, were beyond hope. When he later learned of the success of the evangelical mission at Tierra del Fuego, Darwin recanted and, in a letter dated June 30, 1870, wrote: "It is most wonderful and shames me, as I prophesied utter failure.

---

25 *Descent of Man*, 137.
26 *Life and Letters*, 1:285. Italics not in the original.

**SATIRISTS** • Charles Darwin's theory of evolutionism became a favourite focus for cartoonists and illustrators.

It is a grand success. I shall feel proud if your committee think fit to elect me an honorary member of the society."[27] Furthermore Darwin, who always supported the spread of Western civilization,

27 *Life and Letters*, 2:307.

sent money to support the South American Missionary Society and subscribed to its missionary journal.

*Descent of Man*, as one would expect, brought forth a wide spectrum of responses. As with *Origin*, publisher John Murray invited Whitwell Elwin, the former editor of the *Quarterly Review*, to pass judgement on the book. This country cleric's opposition to *On the Origin of Species*[28] paled in comparison to his total disdain towards *Descent of Man*. He responded: "The arguments in the sheets you have sent me appear to be little better than drivel."[29] In spite of some negative reaction, as with *Origin*, Murray went ahead and published *Descent of Man*.

Within the Darwin household, there had been mixed reactions to *Descent of Man*. Charles had enlisted the services of his wife Emma and Henrietta, his daughter, to be proofreaders. His eighteen-year-old who leaned toward atheism was delighted with the final product but Emma was more tentative. "I think it will be very interesting, but that I shall dislike it very much as again putting God further off."[30]

St. George Mivart (1827-1900) was an impassioned critic. He differed from most in that he was a committed evolutionist and also a confirmed Roman Catholic. Mivart argued that Natural Selection alone could not explain the development of life, especially humanity. He vociferously argued that a Creator must have directed the evolutionary process. Darwin's deistic beliefs revolted against Mivart's contention that God was, and always had been, superintending every facet of the created order. Just a few weeks before *Descent* was published, Mivart's *On the Genesis of Species*, appeared in the bookstores. It was designed to be "a pre-emptive strike on the *Descent of Man* and, coming from a man close to the

---

28 See page 106.

29 Janet Browne, *Charles Darwin: The Power of Place* (New York: Knopf, 2002), 2:453.

30 Henrietta Litchfield, ed., *Emma Darwin: A Century of Family Letters, 1792-1896*, 2 vols. (London: John Murray, 1915), 2:196.

inner circle, it left Darwin badly 'shaken.' He was so angry he could barely speak."[31] Charles wore his religious sentiments very close to his heart and they were easily agitated and frustrated.

Francis Orpen Morris (1810-1893), an evangelical by belief, brought a different but nonetheless ardent antipathy to Darwinism. Like Charles Darwin, he was a born naturalist. As a young boy, he was intrigued with all forms of wildlife. After graduating from Oxford in 1833, he entered the Anglican ministry and served as a lifelong cleric in the parishes of Yorkshire.

His popularity as a writer of natural history soared through his two well-illustrated books: *A Natural History of British Birds* (1850-1857) and *A Natural History of British Butterflies* (1852). He was also well known as "a frequent and voluminous correspondent of the *London Times* and an irrepressible pamphleteer."[32] With the tenacity of a British bulldog, he turned his pen against feminism and cruelty to animals, but it was against evolutionism that he was most strident.

Morris' *All the Articles of the Darwin Faith* "parodied the Anglican creed by beginning every sentence with the phrase 'I believe...'"[33] His ridicule was targeted against *Descent of Man*:

> I [Charles Darwin] believe, I am ready to believe, any-thing—like the Infidel of old—provided only it is not in the Bible....
>
> I believe that the 'imperfection of the geological record', showing no regular chain of species, and giving no proof of my theory, and therefore the most obvious and gravest objection...
>
> I believe that if ever there was such a person as Moses, the five books called the five books of Moses were none

31 Desmond and Moore, *Darwin: The Life of a Tormented Evolutionist*, 577.

32 Charles A. Kofoid, "Francis Orpen Morris: Ornithologist and Anti-Darwinist" in *The Auk*, Vol. 55, No. 3 (July 1938), 497.

33 Browne, *Charles Darwin: The Power of Place*, 2:452.

of his at all, but a mere compilation of some imposter or victim of delusion....

I believe that an assertion 'not proven' is as good as, or better than, one that is proved. Time will tell.[34]

After reading Morris' book, Darwin wrote on the outside, "Keep as a curiosity of abuse."

Rev. Morris' acceptance and esteem within the scientific community is seen best when he was presented with the opportunity in 1877 to read his paper, "A Double Dilemma of Darwinism" before the British Association for the Advancement of Science.

### ON GOD AND HUMANKIND RESOLVED

In his book, *Charles Darwin and the Problem of Creation*, Neal Gillespie has succinctly stated Darwin's resolution concerning God:

In the final analysis, Darwin found God's relation to the world inexplicable; and a positive science, one that shut God out completely, was the only science that achieved intellectual coherence and moral acceptability.[35]

Darwin was caught between two diametric worldviews: a supernatural one in which he envisaged a deity who was remote, unattached and disinterested in the universe he created, and a naturalistic one where human beings functioned as the master of their own destiny and were accountable to no one. But within this naturalistic worldview, Darwin found an intellectual environment in which he could ponder the effects of the natural processes like Natural Selection without the fear of supernatural intervention. God's presence and power were excluded and, furthermore, unwanted.

---

34 Francis Orpen Morris, *All the Articles of the Darwin Faith* (London: W. Poole, 1875), 7, 13,14,16.

35 Neal C. Gillespie, *Charles Darwin and the Problem of Creation* (Chicago: University of Chicago Press, 1979), 133.

Did this compartmentalization of two diametric worldviews cause Darwin any problems? In considering theological questions, he said they were "a painful experience"[36] and they left him in a constant state of "bewilderment."[37] It is understandable that in his letters he viewed his theology as being in a muddled state. In some ways, he could be compared to an individual who wants to be committed to both Hinduism and Christianity. Since both these worldviews are totally opposite, life for this individual would be an endless series of conflicts, and quite intolerable.

By the age of thirty, Darwin had rejected Paley's supernaturalism and embraced Naturalism in which Natural Selection would be the creative genius rather than God. But even within this naturalistic framework in which he felt he could pursue his scientific endeavours, he was dogged by this question: *Who was responsible for the original design?*

Concerning the origin and destiny of humankind, he did not waver for one moment. In *Descent*, he wrote:

> Thus we have given to man a pedigree of prodigious length, but not, it may be said, of noble quality. The world, it has often been remarked, appears as if it had long been preparing for the advent of man: and this, in one sense is strictly true, for he owes his birth to a long line of progenitors. ... nor need we feel ashamed of it.[38]

What then was Charles Darwin—a theist (more correctly, a deist), an atheist or an agnostic? In 1876, he penned these words to a Dutch student:

> I may say that the impossibility of conceiving that this grand and wondrous universe, with our conscious selves, arose

---

36 *Life and Letters*, 1:274.
37 *Life and Letters*, 1:274.
38 *Descent of Man*, 171.

through chance, seems to me the chief argument for the existence of God; but whether this is an argument of real value, I have never been able to decide.[39]

The scientist turned evolutionist could not come to a firm decision in his own mind. Such vacillation has, as one might expect, provided Darwinian scholars considerable diversity of opinion regarding Darwin's religious position. Perhaps, it is best to conclude by saying that the father of modern evolutionism was a 'muddled religionist.'

---

39 *Life and Letters*, 1:276.

## Chapter 8

# On suffering and death

On April 12, 1861 at 4:30 a.m., the Confederate forces bombarded Fort Sumter, South Carolina, and ushered in the American Civil War (1861–1865), the deadliest conflict in the history of that country. 625,000 soldiers perished along with untold civilian casualties.[1] The Confederate cannonade on that fateful morning reverberated around the world. News of this national struggle and its four years of subsequent carnage became international news, particularly in Britain. Even in the correspondence between Charles Darwin and Asa Gray, the progress of the war was followed with intense interest.

### WHAT CAUSES PAIN AND SUFFERING?

A few weeks after the Fort Sumter incident, Darwin wrote to his American botanist-friend and did not mince words on his attitude

---

1  John W. Chambers, II, ed., *The Oxford Companion to American Military History* (Oxford: Oxford University Press, 1999), 849.

concerning the war:

> But I suppose you are too overwhelmed with public affairs
> to care for science.—I never knew the newspapers so
> profoundly interesting. ....Some few & I am one, even
> wish to God, though at the loss of millions of lives, that the
> North would proclaim a crusade against Slavery. In the
> long run, a million horrid deaths would be ample repaid
> in the cause of humanity. ...how I should like to see that
> *greatest curse on Earth* Slavery abolished.[2]

Darwin's animosity toward slavery only intensified as he became
older. Having never forgotten personally witnessing its inhumanity
in South America during the *Beagle* voyage, he was delighted when
the northern states, under the leadership of President Abraham
Lincoln, waged war against this despicable human institution. Emma
Darwin—always a true-blue Wedgwood—became so enraged with
the pro-Confederate bias of the London *Times* that she cancelled
their subscription.[3]

On April 17, 1865, some eight days after the Confederate forces
under General Lee surrendered at Appomattox, Darwin made this
startling statement to Gray: "One thing I have always thought, that
the destruction of Slavery would be well worth a dozen more
years war."[4] In light of the heartache and grief that the American
people had endured, Darwin's remark seemed callous but it truly
revealed his intense antipathy towards slavery. The ravages of the
war, over the past four years in America, could easily be defended.
The goal had been achieved: slavery—with its perverse treatment
of fellow human beings—was to be no more!

Justification for the Civil War, from Darwin's perspective, was
understandable but how could he explain the pain and suffering

---

2   *Correspondence*, 9:163. Italics not in the original.

3   *Correspondence*, 13:223, footnote 7.

4   *Correspondence*, 13:126.

that was an ever-present reality within the natural world? Charles had received much personal satisfaction and joy in discovering the mysteries of life. But coexisting with the beauty and the harmony in nature was the ubiquity of cruelty and carnage. His son, Francis (1848-1925), noted: "Something has already been said of my father's strong feeling with regard to suffering both in man and beast. It was indeed one of the strongest feelings in his nature."[5]

The reality of suffering was impressed upon Darwin's mind as a student at Cambridge. Like many of his fellow classmates, Charles had considered hunting to be a favourite pastime. Once while hunting, this aspiring cleric came upon a bird that had been shot a day or so before. The sight of this senseless act caused Darwin to hang up his rifle for the rest of his life.[6]

This example is but one of myriads within the life span of any individual. Where was a beneficent and omnipotent God? Did he design "Icheumonidae [wasps] with the express intention of their feeding within the bodies of caterpillars or that a cat should play with mice?"[7]

Darwin rejected the biblical teaching that, through Adam's wilful act of disobedience, sin entered the world bringing suffering and death.[8] Consequently, he was forced to find a naturalistic solution. In this regard, his reply to Mary E. Boole (1832-1916) is pertinent:

> I may, however remark that it has always appeared to me more satisfactory to look at the immense amount of pain and suffering in this world as the inevitable result of the natural sequence of events, i.e., general laws, rather than from the direct intervention of God, though I am aware this is not logical with reference to an omniscient Deity.[9]

---

5  *Life and Letters*, 2:377.
6  *Life and Letters*, 1:142.
7  *Life and Letters*, 2:105.
8  See Romans 5:12-14; Genesis 3:1-24.
9  *Life and Letters*, 2:247.

Our early progenitors, according to Darwin, showed great acts of kindness and compassion to their own. But they were "indifferent to the suffering of strangers, or even delight in witnessing them."[10] They extended a similar cruelty to animals as well. Only through the power of Natural Selection could these 'savages' be granted the intelligence and religious values to overcome their natural tendencies.

The relentless struggle for existence has been the natural experience for all living things; they are engaged "in a war dominated by the scarcity of resources and the potential extinction."[11] Natural Selection used suffering and, eventually, death to weed out the weaker species in order that the strong could survive and flourish.

Darwin, by attributing the seemingly wasteful acts and imperfections to Natural Selection, provided God with "divine exoneration"[12]—God was now off the hook! Natural Selection, a process that God, a distant and disinterested deity, initiated, was the culprit. It alone was responsible for all the misery and grief in the world. Thus, suffering was and will always be an ineluctable reality. Even in spite of its ubiquity, pain and suffering should not be tolerated nor willingly accepted, but they are to be conquered solely by the strength of human will and character as there is no divine assistance within the universe.

## THE MEANING OF DEATH

The name 'Charles Robert Darwin' was inscribed on an honorary degree of Doctor of Laws (LL.D.) from Cambridge University. For Dr. Darwin, Saturday, November 17, 1877, the day on which the degree was conferred upon him by his *alma mater*, became the most memorable highlight of his academic career. The newspapers did

---

10 *Descent of Man*, 122.

11 Adam Phillips, *Darwin's Worms: On Life Stories and Death Stories* (London: Faber & Faber, 1999), 8.

12 Neal C. Gillespie, *Charles Darwin and the Problem of Creation* (Chicago: University of Chicago Press, 1979), 127.

not miss the irony as it related to this accolade. The editor of the Anglican journal, *Rock*, commented: "No doubt the affair has its ludicrous side, though a believer will scarcely regard the honour paid to the apostle of evolution as by any means a laughing matter."[13]

Darwin had begun writing his *Autobiography* the year before, on Sunday, May 28, 1876 (finished two months later), as a personal memoir for his family. Published posthumously, it covered his early childhood, university days, life on the *Beagle* and closed with a description of his books.

His first mention of death, as one would expect, was that of his mother. Concerning her death, he had little recollection. He vividly recalled the burial of a soldier of the British Dragoons,[14] which occurred a few weeks after his mother's demise. Since his Shrewsbury school was adjacent to the church's cemetery, Mr. Case, his teacher, took his students, as young as they were, to view this military funeral. The Napoleonic wars were still fresh in the minds of the English. In fact, the officer in command of the ceremony had lost an arm while fighting in France.

In the section of his autobiography where Darwin discussed his religious beliefs, he made reference to the question of immortality. Quoting the physicists of the day, he feared that the sun would burn out unless some catastrophic event intervened. Nevertheless, he opined:

> Believing as I do that man in the distant future will be a far more perfect creature than he now is, it is an intolerable thought that he and all other sentient beings are doomed to complete annihilation after such a long-continued slow progress. To those who fully admit the immortality of the

---

13 As quoted in Janet Browne, *Charles Darwin: The Power of Place* (New York: Knopf, 2002), 2:450.

14 *Autobiography*, 24, footnote 1.

human soul, the destruction of our world will not appear so dreadful.[15]

On February 9, 1865, three days before his fifty-sixth birthday, Darwin—still in shock—wrote to Joseph Hooker on hearing of the death of their mutual friend, paleontologist Hugh Falconer (1808-1865). Following the same line of thought as recorded in his *Autobiography*, Darwin opened the window of his soul even more and revealed his true feelings about the meaninglessness of life.

> My dear Hooker,
>
> I had not heard of poor Falconer's suffering before receiving your note. The thought has quite haunted me since. Poor fellow it is horrid to think of him,—I quite agree how humiliating the slow progress of man is but everyone has his own pet horror and this slow progress or even personal annihilation sinks in my mind into insignificance compared with the idea, or rather I presume certainty, of the sun some day cooling and we all freezing. *To think of the progress of millions of years, with every continent swarming with good and enlightened men all ending in this*; & probably no fresh start until this our own planetary system has again converted into red-hot gas.—*Sic transit gloria mundi*, with a vengence.[16]

Darwin's deistic worldview was responsible for such a pessimistic outlook on life. He had come to believe that the universe, and even he himself, would be annihilated—everything would cease to exist! Is it any wonder that with this attitude Darwin had such an intense aversion and avoidance of the topic of death? As noted earlier, he did not attend the funerals of his father or his beloved daughter, Annie.

---

15 *Autobiography*, 92.

16 *Correspondence*, 13:56. The Latin phrase means: 'So passes away the glory of the world.' Italics not in the original.

In the early months of 1861, Hooker informed Darwin that Professor John Henslow, his mentor at Cambridge, was dying and would like to see him. Darwin wrote to Hooker that he was "not equal to such an exertion."[17] He could not visit Henslow because it would cause him too much mental anguish.

Hooker, Henslow's son-in-law, faithfully informed Darwin of his father-in-law's deteriorating condition. In return correspondence, Darwin always expressed concern for his highly esteemed professor but, on the grounds of his own poor health, he consistently declined to come. In one reply, Darwin commented that he had dined with Thomas Bell (1792–1880) at the Linnean Club in London: "But dining out is such a novelty to me that I enjoyed it. Bell has a real good heart."[18] Incidentally, he did inquire about Henslow but there was no mention of any illness! Two weeks later, Henslow died. Darwin had never arranged a time to visit him.

Moving into the twilight years of his life, Darwin turned his attention to the study of the earthworm. In his book, *The Formation of Vegetable Mould, through the Action of Worms, with Observations on Their Habits*, the first sentence of his conclusion began with this contention: "Worms have played a more important part in the history of the world than most persons would at first suppose."[19] This, Darwin's final book, was published in October 1881 and proved to be a phenomenal success, with thousands being sold in the first week.[20]

His investigation of this lowly creature evinced the truly scientific mind that Darwin possessed. His research concerning the habits of the earthworm indicated their importance in maintaining the

---

17 *Correspondence*, 9:69.

18 *Correspondence*, 9:100. Thomas Bell was the president of the Linnean Club.

19 Charles Darwin, *The Formation of Vegetable Mould, through the Action of Worms, with Observations on Their Habits*, Vol. 28 in *The Works of Charles Darwin*, Paul H. Barrett and R.B. Freeman, eds. (London: Pickering, 1988), 28:136. Hereafter, it will be cited as *Action of Worms*.

20 A. Desmond and J. Moore, *Darwin: The Life of a Tormented Evolutionist* (New York: Warner, 1991), 658.

conditions for the survival of all living things. As witnessed by the sales, his work—much of which was original—caught the interest of the general public. But why at this time of his life would a scientist of Darwin's stature be interested in earthworms? By not being forthcoming about his intentions, Darwin left the door wide open for scholars to speculate—and speculate they have!

Ralph Colp (1924- ), an American psychiatrist, has conjectured that Darwin did indeed have an ulterior motive for choosing earthworms for his concluding major research project. He maintained that the father of modern evolutionism envisioned that these little creatures were pivotal in securing for him a naturalistic afterlife. Consistent with his deistic theology, Darwin believed that "his remains will move through the bodies of worms and perform some useful functions."[21]

In the same vein, Adam Phillips (1954- ), author of *Darwin's Worms*, arrived at the same conclusion. For him, Charles Darwin's abandonment of a personal and immanent God begs an answer to this question: What is the basis for immortality? This British psychotherapist feels that the only valid explanation is found in analyzing the last paragraph of Darwin's seminal work on earthworms:

> The plough is one of the most ancient and most valuable of man's inventions; but long before he existed the land was in fact regularly ploughed, and continues to be thus ploughed by earthworms. It may be doubted whether there are many other animals which have played so important a part in the history of the world, as these lowly organised creatures.[22]

In reality, the seemingly insignificant earthworm, Phillips contends, provides the theological key to the enigma concerning human

---

21 Ralph Colp, "The Evolution of Charles Darwin's Thoughts About Death," *Journal of Thanatology* 3 (1975), 201.

22 *Action of Worms*, 139.

MAN · IS · BVT · A · WORM ·

**DARWIN'S FINAL BOOK** • In 1882, 'Man is but a worm' was printed in *Punch*, a British satirical magazine, in response to Darwin's book on worms.

immortality. "It is as though the earth is reborn again and again, passing through the bodies of worms. Darwin had replaced a creation myth with a secular maintenance myth."[23] Through the digestive process of worms, there has been a continuum in life. Worms "preserve the past, and create the conditions for future

_____

23 Phillips, *Darwin's Worms*, 58.

growth. No deity is required for these reassuring continuities. ... They buried to renew: they digested to restore."[24]

Darwin totally rejected the biblical portrait of humans; they were not created in the image of God.[25] Neither do they possess an innate spiritual nature. For Darwin, spirituality was an acquired capacity that Natural Selection granted to humans as they developed intellectually and culturally. Unlike his wife, he believed that at death came the cessation of life; the only meaningful immortality that any individual could hope for was to be remembered, at least momentarily, by one's family and friends.

In a lengthy article on Darwin's religious life, B.B. Warfield (1851-1921), chairman of Princeton Seminary from 1887 until his death, noted that Charles Darwin had made himself a name which will be remembered for many generations to come. But in evaluating his life from an evangelical perspective, Dr. Warfield wrote:

> And thus, as he approached the end of his long and laborious life, ..."he seemed to recognize the approach of death, and said, 'I am not the least afraid to die.'" And thus he went out into the dark, without God in all his thoughts...[26]

On April 19, 1882 at 4 p.m., Charles Darwin took his last breath at the age of seventy-three. His soul winged its way into eternity.

Following his death, and as a result of pressure placed on British parliamentarians by leading intellectuals, Darwin was given a state funeral. His body was laid to rest, not in the small cemetery at St. Mary's Anglican Church at Downe, but, rather, at Westminster Abbey in London, near the tomb of Sir Issac Newton.

---

24 Phillips, *Darwin's Worms*, 56.

25 Genesis 1:27.

26 B.B. Warfield, "Charles Darwin's Religious Life: A Sketch in Spiritual Biography," *The Presbyterian Review* 9 (1888), 599.

**CHARLES DARWIN** • In his twilight years, Darwin focused his efforts on his studies of the earthworm and writing his *Autobiography*. His death in 1882 was followed by a full state funeral in Britain and he was buried in Westminster Abbey in London.

## Chapter 9
# Deathbed repentance: fabricated or true?

O ver the past twenty years, the author has been approached by individuals who have said that they have either read or heard that, just prior to his death, Charles Darwin became a Christian and completely rejected his evolutionary beliefs. Is there any historical basis for these conjectures?

## LADY HOPE IN AMERICA

In 1915, at a summer Bible conference held from July 30 to August 15 at East Northfield, Massachusetts, an English woman known as Lady Hope (1842-1922) announced at a morning prayer meeting that she had once spent a pleasant afternoon conversing with the renowned Charles Darwin.[1] Her recollections of that visit, even though it had occurred some thirty-four years earlier, were so

---

1  "Lady Hope's Last Words, 1940" in James Moore, *The Darwin Legend* (Grand Rapids: Baker Books, 1994), 132.

vivid and captivating that she was encouraged to write them down for posterity's sake. Before passing judgment on the trustworthiness and historicity of Lady Hope's account, one should at least be familiar with its contents. Lady Hope wrote:

> It was on one of those glorious autumn afternoons, that we sometimes enjoy in England, when I was asked to go in and sit with the well-known professor, Charles Darwin. He was almost bedridden for some months before he died. I used to feel when I saw him that his fine presence would make a grand picture for our Royal Academy; but never did I think so more strongly than on this particular occasion.
>
> He was sitting up in bed, wearing a soft embroidered dressing gown, of rather a rich purple shade.
>
> Propped up by pillows, he was gazing out on a far-stretching scene of woods and cornfields, which glowed in the light of one of those marvellous sunsets which are the beauty of Kent and Surrey. His noble forehead and fine features seemed to be lit up with pleasure as I entered the room.
>
> He waved his hand toward the window as he pointed out the scene beyond, while in the other hand he held an open Bible, which he was always studying.
>
> "What are you reading now?" I asked, as I seated myself by his bedside.
>
> "Hebrews!" he answered— "still Hebrews. 'The Royal Book,' I call it. Isn't it grand?"
>
> Then, placing his finger on certain passages, he commented on them.
>
> I made some allusion to the strong opinions expressed by many persons on the history of the Creation, its grandeur, and then their treatment of the earlier chapters of the Book of Genesis.
>
> He seemed greatly distressed, his fingers twitched nervously, and a look of agony came over his face as he said:

"I was a young man with unformed ideas. I threw out queries, suggestions, wondering all the time over everything; and to my astonishment the ideas took like wildfire. People made a religion of them."

Then he paused, and after a few more sentences on "the holiness of God" and "the grandeur of this Book," looking at the Bible which he was holding tenderly all the time, he suddenly said, "I have a summer house in the garden, which holds about thirty people. It is over there," pointing through the open window. "I want you very much to speak there. I know you read the Bible in the villages. Tomorrow afternoon I should like the servants of the place, some tenants and a few of the neighbours to gather there. Will you speak to them?"

"What shall I speak about?" I asked.

"Christ Jesus!" he replied in a clear, emphatic voice, adding in a lower tone, "and his salvation. Is not that the best theme? And then I want you to sing some hymns with them. You lead on your small instrument, do you not?"

The wonderful look of brightness and animation on his face as he said this I shall never forget, for he added: "If you take the meeting at three o'clock this window will be open, and you will know that I am joining in with the singing."

How I wished that I could have made a picture of the fine old man and his beautiful surroundings on that memorable day!

[At one of the morning prayer services at Northfield, Lady Hope, a consecrated English woman, told the remarkable story printed here. It was afterward repeated from the platform by Dr. A. T. Robertson. At our request Lady Hope wrote the story out for *The Watchman-Examiner*. It will give to the world a new view of Charles Darwin. We should like the story to have the widest publicity. Our exchanges are welcome to the story provided credit is given to *The*

*Watchman-Examiner* and marked copies are sent to us.—
THE EDITOR.][2]

The editor of the *Watchman-Examiner*, "the nation's leading
Baptist magazine,"[3] was present at the Bible conference and he
had immediately recognized the news value of this story. He
approached Lady Hope—a prolific author with "some thirty titles
under her *nom de plume*"[4]—to write an account of this extraordinary event. On August 19, 1915, just four days after the conference
had ended, the story, "Darwin and Christianity," appeared in the
*Watchman-Examiner*.

Little did the editor know the impact and the longevity of this
story. Some eighty years later, it is still being read in the form of a
tract and discussed. Historian James Moore has devoted twenty
years of research—spanning three continents[5]—in order to unravel
the mysteries surrounding the tract. In his book, *The Darwin Legend*,
the most comprehensive study on the topic, he has come to this
conclusion: "Lady Hope was a real person; the story she told has
some historical merit, though it immediately launched a legend."[6]

## AUTHOR OF A LEGEND

Lady Hope was born Elizabeth Reid Cotton on December 9, 1842
in Tasmania. Her father (later known as General Sir Arthur Cotton),
a captain in the British Royal Engineers, returned to England in
1860 after having served abroad. A decade later, he settled with
his family in Dorking, some 24 km (15 miles) from Downe. As
committed evangelical Anglicans, they devoted their lives to the
temperance movement in that area.

---

2  R.C. Newman, "The Darwin Conversion Story: An Update" in *Creation Research Society Quarterly* 29 (Sept. 92), 70-72. Used with permission.

3  Moore, *The Darwin Legend*, 85.

4  Moore, *The Darwin Legend*, 24.

5  Europe, North America and Australia.

6  Moore, *The Darwin Legend*, 26.

At the age of thirty-five, Elizabeth married Admiral Sir James Hope (1808-1881) who at that time was a sixty-nine-year old widower. He shared her passion to proclaim the transforming power of the gospel to those under the grip of alcoholism. Known now as Lady Hope of Carriden, she moved to his Scottish estate and continued her calling. "In pubs and schoolrooms, cottages and castles, she preached and prayed and read the Bible, with drunkards, the destitute and the dying."[7] Even though she had moved away from the church in Dorking, the people were still close to her heart and she sent them frequent reports of her missionary activities.

When her four-year marriage ended in 1881 with the death of her husband, she returned to her family home in Dorking. Lady Hope resumed her gospel and temperance ministries in her home territory. James Moore has speculated that this would have been an ideal time for Darwin to have extended an invitation for her to come to his residence. Furthermore, this British scholar has suggested that such a visit could have occurred some time between September 27 and October 4, 1881.[8]

Dr. Paul Marston, Senior Lecturer in science and history at the University of Central Lancashire, England, also concurred that Lady Hope had indeed met with Charles Darwin.[9] In a very sympathetic article (it provides a balance to Moore's anti-evangelical bias),[10] Marston endeavours to resolve some of the issues concerning Darwin's reference to the book of Hebrews, his urging her to preach the gospel and, finally, the renunciation of his evolutionary theories.[11]

---

7  Moore, *The Darwin Legend*, 85.

8  Moore, *The Darwin Legend*, 166.

9  Paul Marston, "Charles Darwin and Christian Faith." Available online at www.paulmarston.net/papers/scienceandreligion.html (accessed January 14, 2009).

10  "Moore's dislike of evangelicals and fundamentalists is so very apparent that one is left wondering whether this has affected his objectivity in dealing with the story of the evangelical Lady Hope," Malcolm Bowden, *True Science Agrees with the Bible* (Kent: Sovereign Publications, 1998), 271.

11  Marston, "Charles Darwin and Christian Faith."

But why would Darwin, a man of international fame, be interested in associating with this female evangelist? First, Charles was a noted humanitarian; he personally witnessed the destructiveness of alcoholism in the Downe area and would have been very supportive of Lady Hope's efforts in curtailing it. Second, Darwin enjoyed hobnobbing with people of influence. Lady Hope definitely could be categorized as such a person. Through her parents, especially her father, and also her late husband, she was well-connected socially. Through her gospel ministry, she had befriended two of the most well-known American celebrities in Britain at that time, evangelist Dwight L. Moody (1837-1899), and hymnist Ira D. Sankey (1840-1908).

James Moore has offered yet a third reason. On Wednesday, September 28, 1881, Edward Aveling (1849-1898), a young anatomist and partner to Karl Marx's daughter Eleanor, along with physiologist, Ludwig Büchner (1824-1899), were invited by Darwin to his estate for lunch. Since both these staunch atheists were attending the International Congress of the Federation of Freethinkers in London, they took the opportunity to meet with the father of modern evolutionism whom they greatly admired.

After finishing the noon meal, they continued their conversation in Darwin's study. Without any hesitation, it was Charles who alluded to the topic of religion:

> "Why do you call yourselves Atheists?"…"I am with you in thought, but I should prefer the word Agnostic to the word Atheist."
>
> Upon this suggestion was made that, after all, 'Agnostic' was but 'Atheist' writ respectable and 'Atheist' was only 'Agnostic' writ aggressive…. At this he smiled and asked: "Why should you be so aggressive? Is anything gained by trying to force these ideas upon the mass of mankind? It is

**EMMA WEDGWOOD DARWIN** • Emma, pictured here at age seventy-three, was a 'biblical' Unitarian.

all very well for educated, cultured, thoughtful people; but are the masses yet ripe for it?"[12]

As a 'biblical' Unitarian, Emma Darwin disliked the militant atheism of these two German guests. Thus, as Moore has conjectured, Charles, endeavouring "to molify"[13] Emma, wanted to show her that he too could not sympathize with individuals who denied the existence of God. Thus, he extended an invitation to Lady Hope whose religious views were more in keeping with the belief system of the Darwin household.

In 1893, Lady Hope married T. A. Denny (1818-1909), a wealthy Irish businessman. She never relinquished her name from her first marriage. Unfortunately, sixteen years later, in 1909, she became a widow again. Due to financial problems, she left England in 1911 for the United States.

Four years later, Lady Hope attended the Bible conference in East Northfield, Massachusetts, where our story began. Did she go with the intent of telling her story about Darwin? Regardless, it was a story tailor-made for an American audience. Once it was made public, it spread like a forest fire. Darwinism had already taken root in the United States, particularly in the educational system. In 1925, evolutionism would go on trial in Dayton, Tennessee, and the Scopes 'Monkey' Trial would become headline news throughout the nation.

Lady Hope went to this conference in 1915 with a heavy heart, as she had been diagnosed with cancer. Leaving the United States in 1922, she travelled to Sydney, Australia, where she died later that year.

Through the indefatigable efforts of James Moore, Lady Hope's visit with Darwin now seems entirely plausible. But there is still one important complication. In a letter written a few months be-

---

12 Edward Aveling, "The Religious Views of Charles Darwin" (London: Free Thought Publishing Company, 1883), pamphlet.

13 Moore, *The Darwin Legend*, 167.

fore his death, James Fegan (1852-1925), a contemporary of Lady Hope, made this startling revelation: "When Sir Francis Darwin [Charles' third son] says that Lady Hope's story is *a fabrication,* that denial is quite enough for anybody who knows the high standards of truth the Darwins inherited from their father."[14] Known as a man of impeccable integrity and steadfast commitment to the cause of Christ, Fegan laboured in boys' orphanages and in preaching the gospel.

But scholars have been critical of Fegan's unquestioning trust in the Darwin family. Was it not they who censored Darwin's letters and autobiography in order to present to the British public the sanitized and sanctified Darwin of their choosing?

It was in 1990, while doing research for the first edition of this book, that the author was first exposed to the Lady Hope contoversy. Since then, much has come to light about Lady Hope's life and her tract. But there still are a few pieces of this historical puzzle yet to be uncovered.

### DID DARWIN RECANT AND REPENT?

It should be noted that the Lady Hope tract never stated that Darwin had abandoned his evolutionary ideas or that he had made a personal commitment to Jesus Christ. Since most people are unfamiliar with its content, the later embellishments were unavoidable, even predictable.

The available evidence, as seen thorough Darwin's letters during the last years of his life, shows little difference in tone and conviction from those of twenty years earlier. Contrary to the tract's declaration that Darwin was committed to the Bible as the inspired Word of God, he wrote in 1879 to a German student that he "did not believe that there ever had been any revelation."[15] Furthermore, on February 28, 1882—seven weeks prior to his death—he wrote to Daniel Mackintosh that life from non-life might well have oc-

---

14 Moore, *The Darwin Legend,* 156. Author's italics.
15 *Life and Letters,* 1:277.

curred and that proof of God's existence from "so-called laws of nature (i.e., fixed sequence of events) was a perplexing subject, on which he often had thought but could not see his way clearly."[16]

Henrietta Litchfield, Darwin's daughter, came home from London on March 3, 1882 to be with her father. "From then on she closely observed her father's condition."[17] It was she who stated in *The Christian* magazine on February 23, 1922:

> I was present at his [Darwin's] deathbed. Lady Hope was not present during his last illness or any illness. I believe he never ever saw her but in any case she had no influence over him in any department of thought and belief. He never recanted any of his scientific views, either then or earlier. We think that story of his conversion was fabricated in U.S.A.[18]

Not everyone is in agreement with Darwin's daughter's assessment. "In the months before his death there are ample instances indicating his turning to Christianity,"[19] writes Laurence Croft. He maintained that this conversion experience was the result of the testimony of Lady Hope during the last months of Charles' life.

The basis of his argument is a series of inferences. He postulated that Darwin came to the realization that Natural Selection could not be applicable to human beings and as a result he began to moderate his views. This was the first step in resolving an inner conflict in which "one might have expected him indeed to revert

---

16 *More Letters*, 2:171.

17 Ralph Colp, *To Be an Invalid* (Chicago: University of Chicago Press, 1977), 93.

18 Wilbert H. Rusch, "Darwin's Last Hours" in Wilbert H. Rusch, John W. Klotz, Emmett L. Williams, eds, *Did Charles Darwin Become a Christian?* (Georgia: Creation Research Society Books, 1988), 20f.

19 Laurence R. Croft, *The Life and Death of Charles Darwin* (Lancashire, England: Elmwood, 1989), 105. In a similar vein, George Dorsey noted that Darwin died a Christian gentleman. Dr. Dorsey quickly qualified this statement by saying that Darwin never made a profession of faith but that "his life was Christlike, and what a Christian's life is supposed to be and so seldom is" [George A. Dorsey, *The Evolution of Charles Darwin* (N.Y.: Doubleday and Page, 1927), 257].

to Christianity."[20] Paul Marston, a fellow evangelical, made this apt comment: "Croft seems to clutch at straws to reach impossible conclusions about an eleventh hour conversion."[21]

Adrian Desmond and James Moore took a different tack in their biography of Darwin. They realized that he had never professed to be a Christian. These two Darwinian scholars who examined the latter years of Darwin's life focused on the question that continually dogged him: Did God really exist? Consequently, they titled their final chapter, "An Agnostic in the Abbey."[22] Insofar as Emma was concerned, they stated: "Her Christianity was a simple evangelical prescription to gain everlasting life by believing in Jesus."[23]

In an internet article, Dr. Robert Kofal disagrees with the two aforementioned positions. He maintains that neither Charles nor Emma were ever Christians. He came to this conclusion after examining the correspondence between the two of them during the sickness and subsequent death of their beloved daughter Annie in 1851. Their hopelessness spoke volumes to him about their lack of a biblical faith. In their letters "there was absolutely no reference to any thought about life after death for little Annie, nor of ever seeing her again."[24]

## A WORD OF CAUTION

A final word of caution should be made regarding the continued distribution of the 'Lady Hope' tract. It would be a spectacular revelation if the 'father of evolutionism,' in his last days, had indeed become a follower of Jesus Christ and rejected his belief system— a system that has become foundational to all modern learning. But,

---

20 Croft, *The Life and Death of Charles Darwin*, 107.

21 Marston, "Charles Darwin and Christian Faith."

22 A. Desmond and J. Moore, *Darwin: The Life of a Tormented Evolutionist* (New York: Warner, 1991), 664–677.

23 Desmond and Moore, *Darwin: The Life of a Tormented Evolutionist*, 281.

24 Robert E. Kofal, "Charles Darwin: Influences on the Man, His Science, and His Theory," (1996), on-line at www.creationscienceoc.org/Articles/CharlesDarwin.html (accessed January 14, 2009).

as we have seen, the evidence of either of these being a reality is highly questionable.

# Conclusion

Charles Darwin's religious life can be viewed as a four-part drama. The three eternal questions provided the framework for the script.[1]

'Act 1' centered on Darwin's life when he accepted 'biblical' Unitarianism, which he adopted from the Wedgwood influence in his life. This belief system embraced a non-trinitarian God and the Bible as a book of high moral standards.

An 'Interlude' occurred during the time of Darwin's *Beagle* trip around the world. It was during this time that he began to question the Unitarian concept of God and the Bible. Reading the writings of Charles Lyell, Charles was exposed to the possibility of a deistic worldview.

'Act 2' began some time between 1837 and 1839 when Charles made a wholehearted conversion to Deism. He embraced the

---

1  See author's Preface.

**JAMES FEGAN (1852-1925)** • Fegan's social and evangelistic work in the town of Downe did not go unnoticed by the Darwin family.

belief in an impersonal and remote Supreme Being or First Cause. The Bible was soundly rejected as merely a book of myths. His commitment to his wife, Emma, and his children was of primary importance. When the opportunity availed itself, Charles was more than willing to assist the needy within his community. But he felt that his greatest legacy for humanity would be through his scientific endeavours.

'Act 3' recorded Darwin's coming to terms with death. He believed that at death—an unavoidable reality—came the final curtain. Once it dropped, life ceased to be.

The irony of this drama is that, even though Darwin had pushed God out of his everyday world, God was still there. The most eminent scientist of the nineteenth century had the unique opportunity to see him working in the life of one James Fegan. This Brethren layman, feeling called of God to care for the underprivileged boys of London, established his first orphanage in 1872. Since his parents had moved to Downe, he had an opportunity to come in contact with the Darwin family.

Wanting to begin an evangelistic outreach in Downe, and needing a place to conduct these meetings, Mr. Fegan, in a daring move, wrote to Darwin, a man of international fame and of a much higher social standing. He asked permission to use the old school house that Darwin had rented from Sir John Lubbock (1834–1913) and converted into a 'Reading Room' for the villagers. Darwin's reply was:

> You ought not to have to write to me for permission to use the Reading Room. You have far more right to it than we have, for your services have done more for the village in a few months than all our efforts for many years. We have never been able to *reclaim a drunkard* but through your services I do not know that there is a drunkard left in the village. Now may I have the pleasure of handing the Reading Room over to you.[2]

---

2  Henry Pickering, *Chief Men Among the Brethren* (London: Pickering and Inglis, 1931), 191. Italics not in the original.

Mrs. Henrietta Litchfield, Darwin's daughter, in compiling her mother's letters, added this footnote to one dated February 1881: "Old M was a notable drunkard in the village of Downe, converted by Mr. Fegan."[3]

Such demonstration of God's power in reclaiming those caught in the grip of alcoholism had a profound effect on the Darwin family. When Mr. Fegan was conducting his evangelistic services in the Reading Room, which was called 'The Gospel Room,' the meal schedule at the Darwin home was altered so that everyone could attend. There is no record that Charles ever went.

But the God that Darwin had purged from the realm of nature and had pushed back into some distant past, made his presence known in Darwin's own household. Joseph Parslow, his faithful butler for some forty years, and Mrs. Sales, the housekeeper, committed their lives to Jesus Christ as a result of Mr. Fegan's preaching.[4]

The reality of God's redeeming power as witnessed by Charles Darwin in the work of the South American Missionary Society in Tierra del Fuego and in the effective proclamation of the gospel to the alcoholics of Downe and members of his own household seemed to have had little impact on his life. But God was indeed at work in Darwin's world! As foretold in the Scriptures, God would manifest his presence and power in the lives of those who by faith had committed themselves to him through Jesus Christ. To the church in Corinth, the apostle Paul wrote:

> But we have this treasure in jars of clay to show that this all-surpassing power is from God and not from us.[5]

---

3 Henrietta Litchfield, ed., *Emma Darwin: A Century of Family Letters, 1792-1896* (London: John Murray, 1915), 2:244.

4 Pickering, *Chief Men Among the Brethren*, 192.

5 2 Corinthians 4:7 (NIV).

## *Appendix*
# Today's university students and Darwin

For a period of thirteen months, from March 28, 2007 to April 30, 2008, 350 students at the University of Western Ontario (UWO) in London, Ontario, Canada, were randomly approached and asked to answer a questionnaire concerning Charles Darwin's religious views.

## RESULTS OF THE DARWIN QUESTIONNAIRE
### RELIGIOUS PERSUASION

| | | | |
|---|---|---|---|
| Not stated | 82 | Hindu | 11 |
| Roman Catholic | 65 | Other | 11 |
| Atheist | 59 | Buddhist | 10 |
| Protestant | 50 | Jewish | 9 |
| Muslim | 30 | Sikh | 3 |
| Evangelical | 18 | Confucianist | 2 |
| | | *Total* | 350 |

**QUESTION 1**

*In your* opinion, *Charles Darwin believed that the process which brought the universe and life into existence was:*

| | |
|---|---|
| Matter-Energy (No God) + evolutionism | 147 |
| Supreme Being or First Cause + evolutionism | 103 |
| God (Bible) + evolutionism | 49 |
| God (Bible) + no evolutionism (only creationism) | 4 |
| Do not know | 47 |
| *Total* | 350 |

**QUESTION 2**

*In your* opinion, *what did Charles Darwin see his purpose of life to be? (more than one answer permissible)*

| | |
|---|---|
| To improve the world through scientific advancement | 243 |
| To eliminate social injustices (e.g. slavery, cruelty to animals) | 74 |
| To worship God (e.g. praying, obeying the Bible) | 23 |
| Life has no purpose or meaning | 22 |
| Do not know | 51 |

**QUESTION 3**

*What,* do you think, *was Darwin's view about what would happen to himself after he died?*

| | |
|---|---|
| Heaven and hell | 75 |
| Heaven and no hell | 25 |
| Cessation of existence (dust, food for worms) | 164 |
| Reincarnation | 15 |
| Do not know | 71 |
| *Total* | 350 |

**YOUR PERSONAL VIEW**

*What,* do you think, *will happen to you after death occurs?*

| | |
|---|---|
| Heaven | 146 |
| Cessation of existence | 105 |
| Reincarnation | 31 |
| Spiritual continuum | 14 |
| Other | 2 |
| Do not know | 52 |
| *Total* | 350 |

**GENERAL OBSERVATIONS**

The respondents to the Darwin questionnaire demonstrated the diversity of religious backgrounds that are represented on a Canadian university campus. It seems somewhat ironic that Canada, known as a Christian country, has 'not stated' as its top response under religious persuasion![1]

The data confirmed a number of the author's suspicions. First, it was not surprising that the majority of students thought that Darwin believed that naturalistic processes, without any divine intervention, were responsible for the origin of the universe and life. *Charles Darwin's Religious Views* addresses this common misconception by showing that Charles Darwin was never an atheist. Throughout his entire life, he never wavered in his conviction that the complexity of life demanded a Supreme Being or First Cause.

Second, as expected, a substantial number recognized that Darwin, being a scientist, wanted to improve the world through the scientific enterprise. However, it was evident that most students had little familiarity with his personal life. Certainly, our education system can take some responsibility for this deficiency as it is more concerned about teaching or, more accurately, indoctrinating the students with evolutionary dogma than with having them learn about Charles Darwin as a historical personality.

The responses to the third question showed a consistency in the students' thinking. Since the majority believed that Darwin was an atheist, it was totally reasonable to them that Darwin would view death as the cessation of existence.

In terms of the students personal views about what happens after death, 42% believed that they were going to heaven, while

---

1   A few weeks prior to beginning this questionnaire, another one in which 1,200 students had participated was completed for the book, *Eternity Before Their Eyes* (London, Ontario: D & I Herbert, 2007). Even though it entailed a larger sampling, the results were strikingly similar. Again the category 'not stated' topped the list with 266 students; it was followed by 258 Roman Catholics. At this point, there was a difference, as the Protestant students were 221 in number. 157 students identified themselves as atheists.

30% indicated that there was no existence after death—and almost 15% answered 'do not know.'[2] This is certainly evidence of a generation that has little hope for the future!

Over the course of this study, it was the author's privilege to interview a number of professors who were visiting from China. One agreed to meet on campus and, typically, she identified herself as an atheist who believed that at death she would cease to exist.[3] When asked if she was afraid to die, she emphatically said that she was. During the conversation, this social science scholar mentioned that she had a daughter at home in China. When asked if she had ever considered what would happen if her young daughter was to die, without a moment's hesitation, she placed her hands over her face in utter anguish—never before had the author experienced such a response.

In the same vein, secularist Philip Kitcher, in *Living with Darwin*, made this startling admission:

> For the benefits religion promises to the faithful are obvious, and obviously important, perhaps most plainly so when people experience deep distress. *Darwin doesn't provide much consolation at a funeral.*[4]

While in a Roman prison, the apostle Paul sent these words of great consolation and hope to the church in Philippi: "For to me, to live is Christ and to die is gain."[5] What a contrast!

---

2  In *Eternity Before Their Eyes*, 48% chose heaven as their final destination while 21% believed in the cessation of existence (98). How does one account for the drop of 6% in the number of those who think that they are going to heaven, while there was a 9% increase in the number who believed in the cessation of existence? A possible explanation could be that, in the Darwin questionnaire, more foreign students were interviewed. Students from China, trained in an atheistic educational system, consistently considered death to end everything.

3  Interview 330 (April 2, 2008).

4  Philip Kitcher, *Living with Darwin: Evolution, Design, and the Future of Faith* (Oxford: Oxford University Press, 2007), 155. Italics not in the original.

5  Philippians 1:21 (NIV).

# Select bibliography

**PRIMARY SOURCES**

Aveling, Edward. "The Religious Views of Charles Darwin." London: Free Thought Publishing Company, 1883.

Bengough, J.W. *Motley: Verses Grave and Gay*. Toronto: William Briggs, 1895.

Chambers, Robert. *Vestiges of the Natural History of Creation* (1844). Reprint of the 1st edition. Introduction by Gavin de Beer. New York: Humanities Press, 1969.

Darwin, Charles. *The Autobiography of Charles Darwin (1809-1882)*. Edited by Nora Barlow. London: Collins, 1958.

Darwin, Charles. *The Correspondence of Charles Darwin*. Edited by Frederick Burkhardt and Sydney Smith. 15 vols. Cambridge: Cambridge University Press, 1985–forthcoming.

Darwin, Charles. *The Descent of Man, and Selection in Relation to Sex*. Vol. 21 in *The Works of Charles Darwin*. Edited by Paul H.

Barrett and R.B. Freeman. London: Pickering, 1988.

Darwin, Charles. *Diary of the Voyage of H.M.S. Beagle*. Edited by Nora Barlow. Vol. 1 in *The Works of Charles Darwin*. Edited by Paul H. Barrett and R.B. Freeman. London: Pickering, 1986.

Darwin, Charles. *The Formation of Vegetable Mould, through the Action of Worms, with Observations on Their Habits*. Vol. 28 in *The Works of Charles Darwin*. Edited by Paul H. Barrett and R.B. Freeman. London: Pickering, 1988.

Darwin, Charles. *The Life and Letters of Charles Darwin*. Edited by Francis Darwin. 2 vols. New York: Basic Books, 1959.

Darwin, Charles. *More Letters of Charles Darwin*. Edited by Francis Darwin. 2 vols. London: John Murray, 1923.

Darwin, Charles. *On the Origin of Species* (1876), Vol. 16 in *The Works of Charles Darwin*. Edited by Paul H. Barrett and R.B. Freeman. London: Pickering, 1988.

Darwin, Erasmus. *Zoonomia; or, the Laws of Organic Life*. 2 vols. London: J. Johnson, 1794.

FitzRoy, Robert and Charles Darwin. "A Letter, Containing Remarks on the Moral State of Tahiti, New Zealand, &c." *South African Christian Recorder* 2, No. 4 (September, 1836).

FitzRoy, Robert. *Narrative of the Surveying Voyages of His Majesty's ships Adventure and Beagle between 1826-1836* (1839).

Newman, Francis W. *Phases of Faith*. New York: Humanities Press, 1970.

Paley, William. *A View of the Evidences of Christianity*. London: The Society for Promoting Christian Knowledge, 1871.

Priestley, Joseph. *Autobiography of Joseph Priestley*. Introduction by Jack Lindsay. Teaneck: Farleigh Dickenson University Press, 1970.

Priestley, Joseph. *An History of the Corruptions of Christianity*. 2 vols. Birmingham: Piercy and Jones, 1782.

von Kotzebue, Otto. *A New Voyage Round the World in the Years: 1823-1826*. 2 vols. N. Israel/Da Capo: Amsterdam/New York, 1967.

# SECONDAY SOURCES
## ARTICLES

Colp, Ralph. "'Confessing a Murder': Darwin's first Revelation about Transmutation." *ISIS* 77 (1986), 9-32.

Colp, Ralph. "The Evolution of Charles Darwin's Thoughts About Death." *Journal of Thanatology* 3 (1975), 191-206.

Desmond, A. "Robert E. Grant: The Social Predicament of a Pre-Darwinian Transmutationist." *Journal of the History of Biology* 17 (1984), 189-223.

Garfinkle, Norton. "Science and Religion in England 1790-1800." *Journal of the History of Ideas* 26 (1955), 376-388.

Gruber, H. and V. Gruber. "The eye of reason: Darwin's development during the *Beagle* voyage." *ISIS* 53 (1962), 186-200.

Lamoureux, D. "Theological Insights from Charles Darwin." *Perspectives on Science and Christian Faith* (2004) [available online at www.asa3.org/asa/pscf/2004/pscf3-04lamoureux.pdf (accessed March 5, 2007)].

Mandelbaum, Maurice. "Darwin's Religious Views." *Journal of the History of Ideas* 19 (1958), 363-378.

Marston, Paul. "Charles Darwin and Christian Faith." Available online at www.paulmarston.net/papers/scienceandreligion.html (accessed January 14, 2009).

Moore, J.R. "Charles Darwin and the Doctrine of Man." *Evangelical Quarterly* 44 (1972), 196-217.

Newman, R.C. "The Darwin Conversion Story: An Update." *Creation Research Society Quarterly* 29 (Sept. 92), 70-72.

Stecher, Robert M. "The Darwin-Innes Letters: The Correspondence of an Evolutionist with His Vicar." *Annals of Science* 17 (1961), 201-258.

Sulloway, Frank. "Darwin's Conversion: The *Beagle* Voyage and its Aftermath." *Journal of the History of Biology* 15 (1982), 325-395.

Warfield, B.B. "Charles Darwin's Religious Life: A Sketch in Spiritual Biography." *The Presbyterian Review* 9 (1888), 569-601.

## BOOKS

Bowden, Malcolm. *True Science Agrees with the Bible*. Kent: Sovereign Publications, 1998.

Brent, Peter. *Charles Darwin: A Man of Enlarged Curiosity*. London: Heineman, 1981.

Browne, Janet. *Charles Darwin: The Power of Place*. New York: Knopf, 2002.

Browne, Janet. *Darwin's Origin of Species: A Biography*. London: Atlantic Books, 2006.

Browne, Janet. *Charles Darwin: Voyaging*. New York: Knopf, 1995.

Burch Brown, Frank. *The Evolution of Darwin's Religious Views*. Macon, Georgia: Mercer University Press, 1986.

Colp, Ralph. *To Be an Invalid*. Chicago: University of Chicago Press, 1977.

Corey, Michael A. *Back to Darwin: The Scientific Case for Deistic Evolution*. Lanham: University Press of America, 1994.

Cosslett, Tess, ed. *Science and Religion in the 19th Century*. Cambridge: Cambridge University Press, 1984.

Croft, Laurence R. *The Life and Death of Charles Darwin*. Lancashire, England: Elmwood, 1989.

Desmond A. and J. Moore. *Darwin: The Life of a Tormented Evolutionist*. New York: Warner, 1991.

Ellegård, Alvar. *Darwin and the General Reader: The Reception of Darwin's Theory of Evolution in the British Periodical Press 1859-1872*. Goteborg: Acta Universitis Gothenburgensis, 1959.

Gillespie, C.C. *Genesis and Geology*. New York: Harper and Row, 1951.

Gribbon, John and Mary. *FitzRoy: The Remarkable Story of Darwin's Captain and the Invention of the Weather Forecast*. New Haven: Yale University Press, 2004.

Herbert, David. *Eternity Before Their Eyes*. London, Ontario: D & I Herbert, 2007.

Herbert, David. *The Faces of Origins: A Historical Survey of the Underlying Assumptions from the Early Church to Postmodernism*. London,

Ontario: D & I Herbert, 2004.

Herbert, Sandra. *Charles Darwin, Geologist*. Ithaca, New York: Cornell University Press, 2005.

Herbert, Sandra. "Darwin the Young Geologist." Edited by David Kohn. *The Darwinian Heritage*. Princeton: Princeton University Press, 1985.

Keynes, Randal. *Annie's Box: Charles Darwin, His Daughter and Human Evolution*. New York: Riverhead Books, 2002.

Kofal, Robert E. "Charles Darwin: Influences on the Man, His Science, and His Theory." 1996. On-line at www.creationscienceoc.org/Articles/CharlesDarwin.html (accessed May 3, 2007).

Livingstone, David N. *Darwin's Forgotten Defenders*. Grand Rapids: Eerdmans, 1987.

Mellersh, H.E.L. *FitzRoy of the Beagle*. London: Rupert Hart-Davis, 1968.

Moore, James R. "Darwin of Down: The Evolutionist as Squarson-Naturalist." Edited by David Kohn. *The Darwinian Heritage*. Princeton: Princeton University Press, 1985.

Moore, James R. "Of love and death: Why Darwin 'gave up Christianity'." *History, Humanity and Evolution*. Cambridge: Cambridge University Press, 1989.

Nichols, Peter. *Evolution's Captain*. New York: Harper Collins, 2003.

Phillips, Adam. *Darwin's Worms: On Life Stories and Death Stories*. London: Faber & Faber, 1999.

Phipps, William E. *Darwin's Religious Odyssey*. Harrisburg: Trinity Press International, 2002.

Pickering, Henry. *Chief Men Among the Brethren*. London: Pickering and Inglis, 1931.

Secord, James A. *Victorian Sensation: The Extraordinary Publication, Reception and Secret Authorship of the* Vestiges of the Natural History of Creation. Chicago: University Press of Chicago, 2000.

Stott, Rebecca. *Darwin and the Barnacle*. London: Faber and Faber, 2003.

Taylor, Ian. *In the Minds of Men: Darwin and the New World Order.* Toronto: TFE Publishing, 1984.

Thomson, Keith S. *HMS* Beagle: *The Story of Darwin's Ship.* New York: Norton, 1995.

Walters S.M. and E.A. Stow. *Darwin's Mentor: John Stevens Henslow, 1796-1861.* Cambridge: Cambridge University Press, 2001.

Wedgwood, B. and H. *The Wedgwood Circle, 1730-1897.* Westfield, N.J.: Eastview, 1980.

# Index

## BOOKS

*Deo Optimo et Maximo Gloria*
To God, best and greatest, be glory

www.joshuapress.com

Printed in the United States
220021BV00001B/2/P

9 781894 400305